Friedrich Schlögl

Probability and Heat

Friedrich Schlögl

Probability and Heat

Fundamentals of Thermostatistics

With 52 Figures

Springer Fachmedien Wiesbaden GmbH

Set by Vieweg, Braunschweig

Bound by W. Langelüddecke, Braunschweig

ISBN 978-3-528-06343-6   ISBN 978-3-663-13977-5 (eBook)
DOI 10.1007/978-3-663-13977-5

Contents

---

* By an asterisc are designated parts of more special interest. They can be skipped by the reader without lost of continuity.

---

* By an asterisc are designated parts of more special interest. They can be skipped by the reader without
lost of continuity.

---

*   By an asterisc are designated parts of more special interest. They can be skipped by the reader without lost of continuity.

Preface and Introduction

This book is based on lectures for graduate students of physics and physical chemistry. Its main aim is to represent the connections between the microdynamics of molecules and quanta with the macroscopic thermodynamics. There are many excellent textbooks on thermodynamics, and the question arises why another one should be written. In no other field of physics we can find as many different possible ways of introducing the theory as in thermodynamics. The reason is not only that here we have to distinguish between a macroscopic and a microscopic description of the same phenomena. The concept of probability in the statistical theory of thermodynamics introduces a fundamentally new element into physics. Between experience and the conventional theories, the use of probability was an intermediate new methodical element which did not occur in the fundamentals of classical physics before the development of thermodynamics. Originally, probability was not used in pure macroscopic thermodynamics, which had developed independently into a rather closed theory based on its own set of postulates. After further development, however, it became possible to connect these postulates with other fields of physics by deducing thermodynamics from statistical mechanics. As a result of this connection it is not surprising that macroscopic thermodynamics is less concrete and less transparent than other fields of classical physics, notwithstanding that, as a rule, understanding it does not require difficult calculus. Such notions as "entropy", "enthalpy", and their relatives, are more abstract and less directly connected with experience or pictures than most of the basic notions in other fields. Even the restriction that we can apply the rather elementary notion "temperature" only to certain kinds of states of a system, does not support the belief that, apart from the subtelities of time averaging, existence of equilibrium, and other conditions required for its definition, temperature is one of the simplest concepts in physics. Only because temperature is associated with a sensual perception, we accept it as familiar to us.

The use of probability in physics requires certain methodological assumptions which, strictly speaking, are not based on experience, but which are necessary to organize experience. As in most cases in which such assumptions are introduced into thermodynamics without clear separation from empirical elements, they are often associated with conceptual difficulties.

A central aim of this book is to overcome these difficulties in a specific way. The book strives for a clear separation of the physical fundamentals of thermodynamics from the general stochastic fundamentals, which in principle do not need to be restricted to their application in physics. This procedure, in a consistent way, is different from that of the usual procedure of textbooks on thermodynamics. It is chosen only because the author believes that it makes things simpler and more transparent in statistical thermodynamics. Indeed, many well-known basic relations of thermodynamics will turn out to be results already of the general stochastic theory. This is the reason for extensive analogies to thermodynamic concepts and connections occurring in modern theories of computer simulation of physical and nonphysical processes. We shall not deal with this field in this book. The above mentioned systematic separation of the different fundamentals, however, will also be useful for understanding the origin of these analogies.

The general stochastic fundamentals comprise more than the mathematical aspects of probability theory. They also include methods of applying probability theory to experience in an appropriate way. Therefore, in the first section of this book, which is devoted to the stochastic fundamentals, we shall also be concerned with elements of information theory.

The main objective of this book is to analyse the connection between thermodynamics and the dynamics on the level of the most detailed description, in the "phase space". Statistics as the foundation of thermodynamics is here understood to be statistics over microstates in phase space. This topical restriction implies that generally we shall not use statistics on a "mesoscopic" level as an ad hoc assumption and as a starting point in its own rights. Due to this restriction of the book we shall not be concerned with the Boltzmann collision equation, with the master equation, with Langevin forces, or with stochastic processes, albeit these methods have proved to be very successful in nonequilibrium thermodynamics. There we refer to expositions existing in the rich literature on the statistical theory of nonequilibrium processes. Mesoscopic statistics will occur only as result of the basic statistics over the microstates. Another restriction is that we consider nonequilibrium processes only if they occur in the regime of "linear thermodynamics". Hence we shall not include the steadily expanding field of modern research of "nonlinear thermodynamics" because this would lead too far.

Although the major concern of the book is the foundation of thermodynamics on the statistical theory, the macroscopic theory will not be represented only as result of the statistics. The macroscopic theory will be developed once again and independently within the pure macroscopic, phenomenological framework. This is done because this framework is an impressive, logically closed system which can be understood without complicated mathematics. The macroscopic theory is moreover a very useful tool for applications which can also be handled without knowledge of the statistical theory. It should be stressed that the simplicity aspired to in the representation of the basic postulates of the phenomenological thermodynamics is not primarily that of logical economy with respect to the independence of basic axioms. It is rather the simplicity of an utmost transparent connection of the basic assumptions with experience on the one hand, and with their consequences on the other hand. Thus the book does not compete with representations of the mathematical axiomatics of phenomenological thermodynamics, which developed into a separate field of research.

The book may differ from the conventional representations in textbooks not only in the previously mentioned striving for a clear separation of methodical from empirical elements but also in other points. This is the case, for instance, with the interpretation of specific heat as statistical measure in chapter 2.3.10, together with the chapters 5.1.2, and 5.2.4. Another point may be the accent placed on the significance of "information gain" and its correlates "availability" and "produced entropy" in thermodynamics. These concepts elucidate some connections more clearly and at times allow for shorter deductions. In the last subsection 5.4, which has rather the character of an appendix, a model system is discussed which at first sight may look a bit far afield from the primary theme of this book. Nevertheless, this subsection deals with two central questions in thermodynamics. The first one is how a mechanical system with reversible dynamics on the microscopic level can exhibit irreversible macroscopic dynamics. The second question is how the motion of such a system can develop a distinctive time scale separation between microscopic and macroscopic processes. These questions cannot be answered analytically in a general way. For the considered model, however, a rigorous solution of these problems is possible.

The book is meant to be primarily a textbook. Therefore most of the examples of special thermodynamic systems are standards in different fields. The number of the examples, however, is restricted. For more applications we refer the reader to the rich literature on thermodynamics. The author hopes that the book, which is concerned above all with basic connections, will be interesting not only for students but also for academic teachers and other scientists who like the structural analysis of fundamentals in physics.

According to the character of a textbook, this book is not intended to demonstrate new results. Nevertheless, the way of the logical deductions, and of the presentations used in this book, as well as the choice of illustrating examples are not only influenced by literature but also by discussions with colleagues and friends. In this respect I should like to mention the Professors *A. Stahl, J. Meixner, R. Bausch, H.-K. Janssen, R. Bessenrodt,* Dr. *E. Schöll,* and Dr. *C. Escher* in Germany, as well as Professor *C. A. Mead* in Minneapolis, and Professor *R. St. Berry* in Chicago. Particular thank is directed to Professor *V. Dohm* for critically reading certain parts of the manuscript and making valuable proposals for improvements. Above all I should like to mention my unforgettable late academic teacher Professor *Richard Becker* in Göttingen, who, now half a century ago, first raised my love for this field. I gratefully acknowledge the help of my niece Dr. *Aenne Hannon,* Houston, Texas, in reducing the linguistic shortcomings of the manuscript I had written in a language which is not my native tongue. Last but not least, I want to thank the publisher, whose suggestion and interest made this book possible.

Friedrich Schlögl

Aachen, December 1988

How to Read this Book

The book is divided into main sections designated by only one serial number, into subsections designated by two numbers, and into chapters designated by three numbers. The first number always marks the main section, the second the subsection, and the third the chapter. Within a subsection, the equations are designated by two numbers, the first corresponding to the chapter, the second to the equation itself. Footnotes and figures are designated by serial numbers running continuously through the whole book.

The reader who is familiar with probability theory may skip subsection 1.1 and use this part only for occasional reference. The same is true with respect to information theory for subsection 1.2. Subsection 1.3, however, is essential for the whole structure of the book because, even at an early stage, it presents important basic relations of thermodynamics which are results of a general statistical theory, independent of underlying physics.

Several chapters and parts of chapters, in particular those which require more advanced knowledge, for instance of quantum mechanics, may be skipped without loss of the logical continuity, both by beginners and by the reader who is mainly interested in the practical application of thermodynamics. These parts are marked by an asterisk in the title. The already mentioned subsection 5.4 is actually an appendix for specially interested readers.

Subsection 2.3 does not rely on the statistical theory, with the exception of some additional excursions which are made and are obvious as such. This part of the book should be readable without knowledge of all the other parts and the author hopes that it can be helpful also for a reader who is only interested in macroscopic thermodynamics.

References with respect to special literature are made in footnotes. They are made only if the subject is not considered to be of textbook standard.

1   General Statistics

In this main section schemes of a general statistical theory not restricted to physics will be developed. They will give rise to fundamental features of the concepts of thermodynamics. We shall start with the contemplation of the concept of probability and shall then be concerned with the definition of information measures. These measures will be helpful when seeking an answer to the question, of how to find adequate probabilities if only relatively few dates are given. This problem can be solved in particular situations typical of thermodynamics and yields results which determine characteristic properties of the basic structure of the thermodynamic theory. Since, however, these results are not restricted to thermodynamics, as already emphasized in the general introduction, a particular intention of the following representation will be, to separate such properties of thermodynamics which are determined by the general statistical theory from those properties which arise from physics, and to avoid a mixing of the different roots of the theory. The physical basis of thermodynamics will enter into later sections of this book. Indeed, in this main section the reader will already get to know a large part of the characteristic relations of thermodynamics.

> **Remark:** As already mentioned in the Introduction, such general relations are valid also in nonphysical applications of the statistical theory. To give more elaborate examples, without going into details, let us mention the recent theories of simulations of complex systems by computer experiments. These theories find application also in economical problems as that of the "traveling salesman". There we can find relations well known in conventional thermodynamics if we replace quantities like energy and temperature with certain quantities of entirely different character. The common origin is the statistical theory.

1.1   Probability

In this subsection we shall be concerned with the theory of probability to an extent sufficient for our purposes. As probabilities belong to events, we shall first discuss those properties of events which in our context are important for the probability theory. After this we shall consider the problem of defining probabilities. The history of the definition of probability is a striking example of the development of a scientific concept from an "explicandum" to a "definitum". It will be outlined at least with the main steps. In the chapters subsequent thereafter we shall delve into the particulars which will be used in this book, like for instance properties of mean values, of moments and cumulants. A more special subject is the "central limit theorem" which will be decued in the last chapter of this subsection. It yields a distinction of the "normal" or *"Gaussian* distribution" that plays a central role in statistics.

1.1.1   Events

Connections between events. To give an illustration, let us consider the special case first that the events are results of experiments. We shall later also introduce events as for instance the dynamical state of all molecules in a gas. The corresponding experiment would require the observation of each molecule in the enormous multitude of say about 10^{23} molecules. Of course, this is completely impossible.

We can connect an event A_1 with another one A_2 by the operation *"or"* and obtain a further event, the event that *"A_1 or A_2"* happens. We call this the *union* of A_1 with A_2, and symbolize it by $A_1 \cup A_2$. Another operation is *"A_1 as well as A_2"*, or in shorter form, the logical *"and"* in *"A_1 and A_2"*. This is called the *intersection* of A_1 with A_2, or the *joint event* of A_1 and A_2, symbolized by $A_1 \cap A_2$. To give an example; it is easier to find a student trained in mathematics *"or"* biology $(A_1 \cup A_2)$ than in mathematics *"and"* biology $(A_1 \cap A_2)$. As will be explained later, there are analogies between the intersection and the product of numbers. Therefore the intersection is likewise called the *product* of A_1 and A_2, and we prefer to write it in the simple form $A_1 A_2$. The two linkage operations union and intersection are commutable:

$$A_1 \cup A_2 = A_2 \cup A_1, \tag{1.1}$$

$$A_1 A_2 = A_2 A_1. \tag{1.2}$$

The result that the event A does not occur is also an event and is called *"not A"* or the *contrary* of A. This is symbolized by $\bar{A}$. It is moreover convenient to introduce the so-called *certain* event S which always occurs. This is the event that one of the possible results of the considered experiment occurs at all, independently of which one in particular. We define also the contrary $\bar{S}$ of S as the event which certainly never occurs. This is called the *zero event* 0. With any event A the following relations hold:

$$A \cup \bar{A} = S, \tag{1.3}$$

$$A \cup S = S, \tag{1.4}$$

$$A \cup 0 = A. \tag{1.5}$$

We can illustrate the operations and relations if we represent the events by hitting a certain target. In Figure 1 each sphere represents an event as target. The hatched areas in case (a) or (b) represent the union or the product respectively. The change from A to $\bar{A}$ is generally obtained by exchanging the hatched area with the blank one.

It should be mentioned that the product is not an independent newly introduced operation if the union and the contrary are already introduced, because

$$A_1 A_2 = \overline{\bar{A}_1 \cup \bar{A}_2}. \tag{1.6}$$

This can easily be seen by means of a picture. Thus in systematic axiomatics, restricted to independent postulates, the introduction of the product is redundant and has to be avoided. The introduction, however is useful if we prefer a system in which the consequences of the basic statements become more conspicuous.

As consequences of eq. (1.6) and the preceding relations we obtain the equations

$$A S = A, \tag{1.7}$$

$$A 0 = 0, \tag{1.8}$$

$$\bar{A} A = 0. \tag{1.9}$$

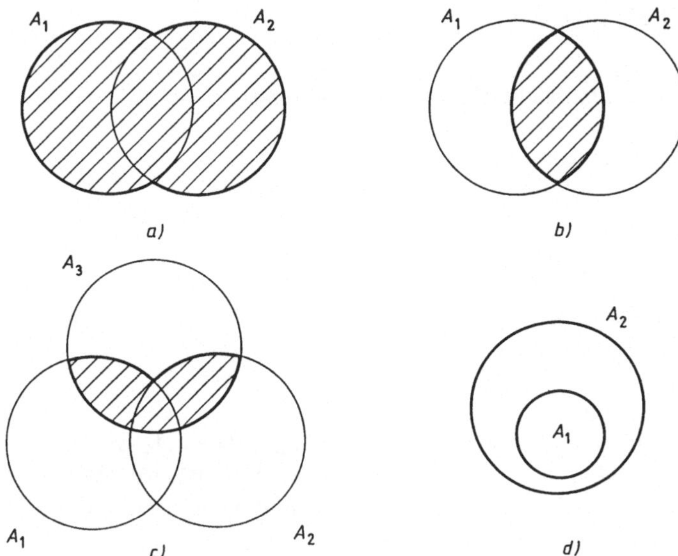

Figure 1 Operations between events. Events are represented as areas. The hatched areas represent: in a) the "union" $A_1 \cup A_2$ of A_1 and A_2, the logical "or", in b) the "intersection" $A_1 \cap A_2$ or "product" $A_1 A_2$, the logical "and". c) demonstrates the distributive law eq. (1.12), and d) the special case that A_1 "induces" A_2 (also expressed by A_2 "includes" A_1).

Analogously, eq. (1.1) leads to eq. (1.2). The *associative law* for the union

$$A_1 \cup (A_2 \cup A_3) = (A_1 \cup A_2) \cup A_3 = A_1 \cup A_2 \cup A_3 \tag{1.10}$$

leads to the *associative law* for the product

$$A_1 (A_2 A_3) = (A_1 A_2) A_3 = A_1 A_2 A_3. \tag{1.11}$$

In both cases it is allowed to drop the brackets as done in the last expression on the right hand side.

Union and product occur both in the *distributive law*

$$(A_1 \cup A_2) A_3 = (A_1 A_3) \cup (A_2 A_3). \tag{1.12}$$

We shall restrict the foundation of this law to the demonstration of the corresponding picture (c) in Figure 1.

With respect to the rules, the product of events is closely analogous to the product of numbers. The union is not analogous to the sum of numbers in the same way. The analogy, however, holds in the special case that the events A_1 and A_2 are *disjoint*. That means, if their product $A_1 A_2$ is zero. Only then the union is called the *sum* of A_1 and A_2:

$$A_1 \cup A_2 = A_1 + A_2 \qquad \text{if } A_1 A_2 = 0. \tag{1.13}$$

The distribution law then takes the form familiar for numbers:

$$(A_1 + A_2) A_3 = A_1 A_3 + A_2 A_3. \tag{1.14}$$

We may write eq. (1.3) in the form

$$A + \bar{A} = S. \tag{1.15}$$

In the following, for events we also shall use the notations

$$A_1 + A_2 + ... + A_n = \sum_{i=1}^{n} A_i, \tag{1.16}$$

$$A_1 A_2 ... A_n = \prod_{i=1}^{n} A_i, \tag{1.17}$$

well known for numbers.

In Figure 1 the special case

$$A_1 A_2 = A_1 \tag{1.18}$$

is represented with picture (d) which means that A_1 *induces* A_2 or: if A_1 occurs, A_2 always occurs as well. We also say that A_2 *includes* A_1. For instance, the event of finding the number twelve in a game induces the event of finding an even number. Finding an even number includes also the number twelve.

Sample sets. The tossing of a die once yields exactly one of six possible results. These six results are an example of a *sample set* or a *complete disjunction*. Generally so is called a set of disjoint events of which certainly one occurs. If $U_1, U_2, ... , U_n$ are these events, this means:

$$U_i U_j = U_i \delta_{ij}, \tag{1.19}$$

$$\sum_{i=1}^{n} U_i = S. \tag{1.20}$$

Here we used the *Kronecker* symbol δ_{ij} which designates 1 if $i = j$, and zero if $i \neq j$. The set of the U_i is called *disjoint* or a *disjunction* for eq. (1.19) and *complete* for eq. (1.20). Thus a sample set is the same as a complete disjoint set of events.

Field of events. A *field of events* is a set of events which comprises with any event A also the contrary $\bar{A}$, and with any pair A, B of events also the union $A \cup B$. The field comprises S and 0 due to eq. (1.3), and comprises AB due to eq. (1.6). We see that all the operations introduced previously never lead out of the field. This is the distinguishing feature of the concept "field".

> **Remark:** The algebraic properties define the field to be a *Bool algebra*. These algebraic properties are the existence of two connection operations, the rules of calculation given above, and the just formulated two field properties.

A sample set is not a field. We can however, always construct a field by use of a sample set $U_1, U_2, ... , U_n$. We define the field as the set of all possible sums of the events U_i. This is the set of the events

$$A = \sum_{i=1}^{n} \eta_i U_i, \tag{1.21}$$

where η_i is equal to zero or to one.

1.1.2 Definitions of Probability

Probability certainly is a very important concept in modern physics and beyond that in modern science. Nevertheless the discussion about the interpretation and the position of probability in science is by no means finished, and has been carried on intensively for more than a century. The definition of probability is a central problem in science and has a long history. It seems worthwhile to go into more details of this question in this book. As we shall explain later in this chapter, there is a system of axioms which defines probability mathematically. I.e., the axioms fix the rules for the calculation with probabilities uniquely. By the application of the concept "probability" in science, however, we are confronted with further questions of semantic character. We need an interpretation which allows us to decide how to connect empirical observations or knowledge of other kind with probabilities.

Generally we can say that a probability P expresses an expectation for the occurrence of an event A under certain conditions. We call these conditions an event c. Usually we compare P of different events A belonging to the same field of events. The higher the value of P, the higher the expectation for the corresponding event. The event c may or may not belong to this field. In an explicit form, this probability is designated by $P(A \mid c)$. In most cases, different A under the same condition c are compared with one another. Then usually c is dropped in the notation and the probability of A is designated simply by $P(A)$. The probability is a real positive number between zero and one:

$$0 \leqslant P(A) \leqslant 1. \tag{2.1}$$

In the following we shall discuss different definitions of the probability which are not contradictory to one another, but rather represent different stages of the development of the concept.

The classical definition. This is the name of the historically oldest proposal of a quantitative definition of probability. As will be explained below, this definition is not satisfactory. Nevertheless it is still living on some representations as "the" definition. Again we use the die as an example and ask for the probability of tossing an even number of dots, called event A. The six different numbers of dots form a sample set and are called the *possible* events. The three even numbers 2, 4, 6 of dots are called the *favorable* events. The ratio of favorable to possible events is defined to be the probability $P(A)$, which is 1/2 in our example.

Generally speaking, the classical definition supposes that the field of all A is based on a sample set of *elementary* events U_i of equal probability. The field is constructed in correspondence to eq. (1.21). All U_i of the sample set are called the *possible* events and the particular U_i, the union of which is A, are called *favorable* events. $P(A)$ is the ratio of the numbers of favorable to the number of possible events.

First, we see that this concept is not applicable to cases in which we cannot find a sample set of "elementary" events. In the last century, for instance, without our modern knowledge of microbiology, it was not possible to decompose the events considered in *Mendel*'s laws about flower colours into elementary events of equal probability. Nevertheless, with these laws the concept of probability was used with success in a precise quantitative way.

Secondly, the anticipating use of *equal probability* in the definition of probability itself makes the definition logically incomplete.

The classical definition was developed and applicable in the theory of games of chance. We, however, need a more general definition in science.

The empirical definition. This is also called the "definition by *v. Mises*". We consider an experiment which can be repeated very often, always under the same conditions. If N is the whole number of these experiments and exactly $N(A)$ of them yield the result A, we call the ratio

$$\frac{N(A)}{N} = h_N(A) \tag{2.2}$$

the *relative frequency* of the result A. A fundamental assumption now is that for increasing N a limit exists

$$P(A) = \lim_{N \to \infty} h_N(A), \tag{2.3}$$

which is called the *probability* of A.

In Figure 2 a possible sequence of $h_N(A)$ as a function of N is drawn as an illustration of the expected qualitative behaviour of $h_N(A)$.

The existence of the limit is a basic postulate. If we find a seeming deviation from the postulate, for instance if $h_N(A)$ after a certain value N_0 of N tends towards another value than before, we should assume that uncontrolled conditions have changed, but not that the postulate is wrong. In this sense the postulate is not an empirical law but a methodical prescription for systematizing experience. The postulate contains the supposition that the connection between the condition c and A is invariant with respect to time shift.

The empirical definition is applicable if we can repeat the experiment as often as we wish. Then the probability is a measurable quantity like any observable in physics. This means that we can find the value of the probability to required bounds of accuracy.

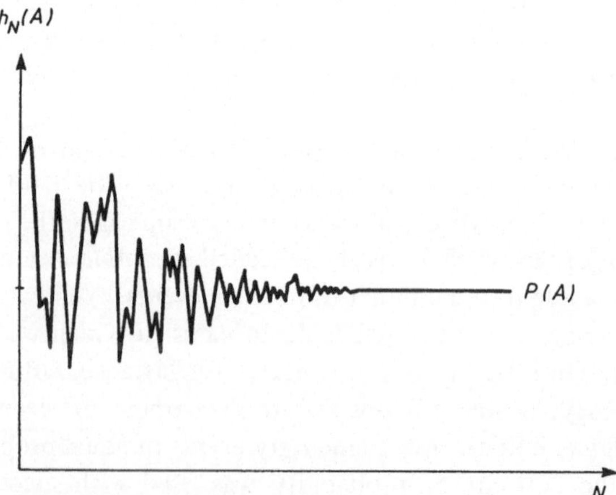

Figure 2 An example of the typical dependence of the relative frequency $h_N(A)$ of an event A on the increasing number N of observations

It should be stressed that vanishing $P(A)$ does not mean that A never can occur. It may be that A sometimes occurs, but that the limit of $h_N(A)$ is zero.

Probability as a measure of expectation. We use the concept of probability also in cases in which we are not able to repeat a corresponding experiment. For instance, we may ask for the probability that a comet will be captured by a planet. It is possible that the answer can be given by a good theory about the dynamics of the objects. Certain conditions given by adequate observations may not allow a unique conclusion with respect to the capture but may allow the calculation of the probability of the capture. In thermodynamics we are usually concerned with such theoretically calculated probabilities and not with observed relative frequences. The situation is different in other fields of experience, as in biology or economics.

It is not necessary for the use of probability $P(A|c)$ that the observation which yields the *conditional event* c be done earlier than the logically *consecutive event* A occurred. We can also ask into the direction of the past. For instance we can ask whether a moon was captured by a planet into an orbit and did not emerge from the planet in whichever form. One of these events happened irrevocably. It is, however, unknown which one. Doubt exists for the observer and the probability is dependent on his knowledge. This knowledge is the conditional event c. This may illustrate what we mean by saying that the probability is a measure of expectation based on the knowledge of the observer. (The probability of different distributions of cards in a card game would be changed for the player if he clandestinely looked into the cards of his neighbour. Yet the card distribution is not changed.)

The definition of probability as the limit of the relative frequency can also be used in cases in which the events are not results of an observation but results of a theory. Let us assume that, due to a good theory, the motion of a mass point in space is well determined and known if the initial position is given. We may divide the part of the space which can be reached by the motion into cells and ask for the event that the mass point is in a particular cell. The relative frequency of this event at a certain time t is the ratio of the sum of the time intervals the mass point was in the cell at t. The limiting value of this ratio for t going to infinity is independent of the starting time and corresponds to the definition of probability by eq. (2.3). It is indeed the probability in case the starting time or the starting point is not known. Then the event is random. If the initial situation is precisely given, we know exactly whether the event, which is no more random, occurs at a certain time or not. The mentioned limit is independent of this knowledge and does not express our expectation in this case. Then the name "probability" has to be taken with care. It than is solely used for a relative frequency. With this meaning the name is common also in connection with the study of deterministic dynamical systems by computer experiments. There the results are uniquely determined by the dynamics and initial conditions. This example may illustrate the possible ambiguity. The events are practically unpredictable for the observer, for whom in fact they are random events and for whom the time limit of the relative frequency is a genuine probability, expressing indeed a mere expectation. This shows once more that the property of an event to be random is not independent of the knowledge of the observer.

The axiomatic definition. We always require that the probability $P(A)$, defined in whatever way, has to fulfil the following relations, which in particular are satisfied by the relative frequencies in case that the experiment can be repeated very often: For all A of a field of events we postulate

$$P(A) \geqslant 0, \tag{2.4}$$

$$P(S) = 1, \tag{2.5}$$

$$P(A_1 + A_2) = P(A_1) + P(A_2). \tag{2.6}$$

As shown by *A. N. Kolmogorov*, these relations give a complete axiomatic definition of probability in mathematics[1]. Strictly speaking, the original system of axioms given by *Kolmogorov* included the definition of the field of events A. Therefore his system was more extensive than eqs. (2.4, 2.5, 2.6). Here we prefer to separate the introduction of events and fields from that of the probability for which the three axioms eqs. (2.4, 2.5, 2.6) are sufficient. Nevertheless we shall call these three relations *Kolmogorov axioms*.

In the following we shall be concerned with deducing relations from these axioms. We see directly

$$P(A) + P(\bar{A}) = 1, \tag{2.7}$$

and thus

$$P(A) \leqslant 1. \tag{2.8}$$

For any two events A_1, A_2 of the field we can separate the union into two disjoint parts by use of the equation

$$A_1 \cup A_2 = A_1 + \bar{A}_1 A_2. \tag{2.9}$$

This separation is illustrated in Figure 3 in which $\bar{A}_1 A_2$ is represented by the hatched area. The figure shows moreover the separation of A_2 into disjoint events

$$A_2 = A_1 A_2 + \bar{A}_1 A_2, \tag{2.10}$$

which also can be obtained formally by eqs. (1.7, 1.14, 1.15). Thus we gain

$$P(A_1 \cup A_2) = P(A_1) + P(\bar{A}_1 A_2), \tag{2.11}$$

$$P(A_2) = P(A_1 A_2) + P(\bar{A}_1 A_2). \tag{2.12}$$

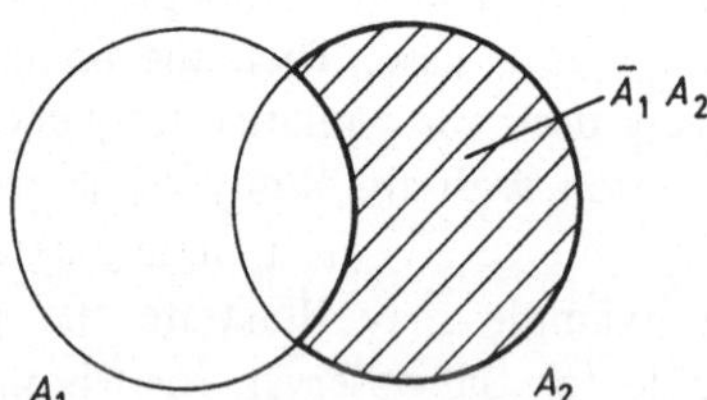

Figure 3

The separation of the events A_1, A_2 into disjoint events A_1 and $\bar{A}_1 A_2$; see eq. (2.9)

---

[1] *A. N. Kolmogorov*, Grundbegriffe der Wahrscheinlichkeitsrechnung (Springer, Berlin 1933); *A. N. Kolmogorov*, Foundations of the Theory of Probability (Chelser Publ. Comp., New York 1950).

Elimination of the last term yields

$$P(A_1 \cup A_2) \leqslant P(A_1) + P(A_2). \tag{2.13}$$

In the special case that A_2 includes A_1 we obtain through eq. (2.10):

$$P(A_1) \leqslant P(A_2) \qquad (\text{if } A_1 A_2 = A_1). \tag{2.14}$$

For a sample set $U_1, \ldots, U_n$ the so-called *normalization relation* holds

$$\sum_{i=1}^{n} P(U_i) = 1. \tag{2.15}$$

For any event A of type eq. (1.21), i.e. of the field constructed out of the sample set, we may write

$$P(A) = \sum_{i=1}^{n} \eta_i P(U_i). \tag{2.16}$$

The whole set of the n probabilities

$$P(U_i) = P_i \tag{2.17}$$

in the following discussion will be called a *probability distribution*. We shall designate the whole set and thus the distribution often by the abbreviation P (generally we shall often abbreviate a whole set of quantities X_i by the corresponding letter X without sub- or superscript).

Conditional probabilities. It was already emphasized that the probability of an event A is defined only with respect to a given conditional event c. As a rule, this is not explicitly expressed in the notation $P(A)$ if c always remains unchanged during the consideration, and thus provides a general framework. If, however, in some cases a further condition as an event B is required, yet in other cases not, it is necessary to distinguish the *conditional probability* $P(A|B)$ of A from the unconditional or *absolute probability* $P(A)$. For instance, the absolute probability of reaching the age of 80 years is different from the conditional probability of reaching this if you are already 70.

In case of the repeated experiment we can define a conditional relative frequency

$$h_{N(B)}(A|B) = \frac{N(AB)}{N(B)}. \tag{2.18}$$

The product AB is the event that A as well as B occur as result. From this equation we arrive at the relation for the conditional probability

$$P(A|B) = \frac{P(AB)}{P(B)}. \tag{2.19}$$

This relation can be used as definition, but is applicable only if $P(B)$ is not zero.

If in particular

$$P(A_2|A_1) = P(A_2), \tag{2.20}$$

the two events A_1, A_2 are called *uncorrelated* with each other. To begin with, this equation reads that A_2 is independent of A_1. As a consequence of this independence, however, the equation holds

$$P(A_1 A_2) = P(A_1) P(A_2), \tag{2.21}$$

which is symmetric in A_1, A_2. Therefore eq. (2.20) is equivalent with

$$P(A_1 | A_2) = P(A_1). \tag{2.22}$$

From this follows that A_1 is independent of A_2 as well. The relation "uncorrelated" is reciprocal. Therefore we may say, that the two events are uncorrelated with "each other".

1.1.3 Random Quantities

Any quantity x which can assume different values $x^{(i)}$ with certain probabilities P_i is called a *random quantity* or *random variable*. The occurrence of a special value $x^{(i)}$ is a random event U_i. All these events U_i form a sample set. Thus for

$$P_i = P(U_i) \tag{3.1}$$

the normalization eq. (2.15) is satisfied

$$\sum_{i=1}^{n} P_i = 1. \tag{3.2}$$

So far we have supposed the possible values $x^{(i)}$ to be discrete. If the possible values x' of the random variable x are continuous, say filling the interval

$$a \leqslant x' \leqslant b, \tag{3.3}$$

then it is adequate to ask for the probability that the value of x lies below x':

$$P(x \leqslant x') = \Phi(x'). \tag{3.4}$$

$\Phi(x')$ is a monotonously increasing function of x'. As a rule, it is called the "distribution function" in mathematical books. In physical literature, however, this name is conventionally used for the *probability density* $\rho(x')$ defined by the requirement that the probability of finding x between x' and $x' + dx'$ be

$$P(x' \leqslant x \leqslant x' + dx') = \Phi(x' + dx') - \Phi(x') = \rho(x') dx'. \tag{3.5}$$

In this book we shall also use this terminology and we shall call the density the *probability distribution* or simply the *distribution*.

We can include the case of discrete values x' (maybe only in certain intervals of x), by use of the *Dirac* delta-function $\delta(x)$. For instance the case of discrete values $x^{(i)}$ described at the beginning can be included by writing

$$\rho(x) = \sum_{i=1}^{n} \delta(x - x^{(i)}) P_i. \tag{3.6}$$

Here and in the following the dash at x' will be dropped. It always holds that

$$\int_a^{x_0} \mathrm{d}x\, \rho(x) = \Phi(x_0), \tag{3.7}$$

and the normalization relation is fulfilled:

$$\int_a^b \mathrm{d}x\, \rho(x) = 1. \tag{3.8}$$

Many dimensions. We shall consider now the case that the random events are determined by not only one random variable x but by a set of d variables, represented by a d-dimensional vector

$$(x_1, x_2, \dots, x_d) = x. \tag{3.9}$$

The events then are mapped in the d-dimensional parameter space x with the volume element

$$\mathrm{d}x_1 \mathrm{d}x_2 \dots \mathrm{d}x_d = \mathrm{d}^d x. \tag{3.10}$$

In this case we define a probability density $\rho(x)$ in this space by requiring that $\rho(x)\,\mathrm{d}^d x$ be the probability of finding the random variables x in the volume element $\mathrm{d}^d x$ at the mapping point x. The probability of finding it in a partial area Δx of the x-space is given by the integral

$$\int_{\Delta x} \mathrm{d}^d x\, \rho(x) \tag{3.11}$$

extended over this partial area. By extension over the whole parameter space, the normalization relation holds:

$$\int \mathrm{d}^d x\, \rho(x) = 1. \tag{3.12}$$

Mean values. For any function $\varphi(x)$ in the parameter space x we can form the so-called *mean value, average* or *expection value* of φ:

$$\langle \varphi \rangle = \int \mathrm{d}^d x\, \rho(x)\, \varphi(x). \tag{3.13}$$

The integration runs over the whole x-space. By the way, eq. (3.12) can be regarded as a special case of eq. (3.13).

The mean value is a special case of a "functional". Generally a functional is the attaching of a number to each function $\varphi(x)$ of a certain mannifold. Such a mannifold often is called a *function-space*. The mean value in particular is a linear functional. This means, that for any two functions of the function-space $\varphi_1(x)$, $\varphi_2(x)$ and for any two numbers c_1, c_2 it always holds that

$$\langle c_1 \varphi_1 + c_2 \varphi_2 \rangle = c_1 \langle \varphi_1 \rangle + c_2 \langle \varphi_2 \rangle. \tag{3.14}$$

Uncorrelated random quantities. We call two random quantities x_1, x_2 *uncorrelated* if the occurence of their possible values are uncorrelated events:

$$\rho(x_1, x_2) = \rho_1(x_1)\,\rho_2(x_2). \tag{3.15}$$

Then

$$\langle x_1 x_2 \rangle = \langle x_1 \rangle \langle x_2 \rangle. \tag{3.16}$$

1.1.4   Moments and Cumulants

In the practical application of statistics we are often confronted with the question how to draw conclusions about the probability distribution $\rho(x)$ of a random variable x from experiments which do not give relative frequencies of the possible values of x but give mean values of certain functions of x. Special mean values of this kind are the so-called *moments* of x. The *moment of order ν* is the mean value

$$M_\nu = \langle x^\nu \rangle \tag{4.1}$$

of x to the power ν. The more moments of increasing order $\nu = 1, 2, \dots$ we know, the more we know about the distribution $\rho(x)$. In most cases we obtain only moments of the very lowest order, say one and two, by more or less direct observations. Therefore the lowest moments are of particular interest in the theory of statistics. All moments can be obtained by the so-called *generating function*

$$Z(\alpha) = \langle \exp(\alpha x) \rangle = \sum_{\nu=0}^{\infty} \frac{\alpha^\nu}{\nu!} M_\nu \tag{4.2}$$

of a parameter α introduced especially for this purpose.

The scheme can be generalized for more than one random variable. For a set of random variables $x_1, x_2, \dots, x_d$ the moments are defined by

$$M_{\nu_1 \nu_2 \dots \nu_d} = \langle x_1^{\nu_1} x_2^{\nu_2} \dots x_d^{\nu_d} \rangle, \tag{4.3}$$

where ν_k is the power of x_k in the product the mean value of which is the moment. The sum of these powers

$$\nu = \nu_1 + \nu_2 + \dots \nu_d \tag{4.4}$$

is called the *order* of the moment. By use of the scalar product notation

$$\alpha_1 x_1 + \alpha_2 x_2 + \dots + \alpha_d x_d = \alpha x \tag{4.5}$$

the generating function can be written

$$Z(\alpha) = \langle \exp \alpha x \rangle = \sum_{\nu_1 \dots \nu_d} \frac{\alpha_1^{\nu_1} \dots \alpha_d^{\nu_d}}{\nu_1! \dots \nu_d!} M_{\nu_1 \dots \nu_d}. \tag{4.6}$$

The moments of the order one are the ordinary mean values.

There is another set of mean values which are certain polynomials of the moments but which have an important property, the moments do not possess. These mean values are called *cumulants*. They are generated by the generating function

$$\Gamma(\alpha) = \ln \langle \exp \alpha x \rangle = \sum_{\nu_1 \dots \nu_d} \frac{\alpha_1^{\nu_1} \dots \alpha_d^{\nu_d}}{\nu_1! \dots \nu_d!} \, C_{\nu_1 \dots \nu_d} \tag{4.7}$$

in an analogous way to that in which the moments are generated by $Z(\alpha)$. This analogy is the reason for designating the cumulants with

$$C_{\nu_1 \dots \nu_d} = \langle x_1^{\nu_1} x_2^{\nu_2} \dots x_d^{\nu_d} \rangle_c \tag{4.8}$$

by association with eq. (4.3). To distinguish them from the moments, the bracket is marked by the subscript c. We may say, that whereas the moments are coefficients in the expansion of $Z(\alpha)$ into an infinite sum, the cumulants are the coefficients in the expansion of the same function into an infinite product. The order ν of a cumulant is defined by eq. (4.5) as it was for moments.

The distinguishing property of the cumulants is that they are additive for uncorrelated random quantities. For two uncorrelated variables x_1, x_2 it holds that

$$Z(\alpha) = \langle \exp \alpha_1 x_1 \rangle \langle \exp \alpha_2 x_2 \rangle. \tag{4.9}$$

Thus $\Gamma(\alpha)$ is additive with respect to x_1, x_2 and

$$\langle (x_1 + x_2)^\nu \rangle_c = \langle x_1^\nu \rangle_c + \langle x_2^\nu \rangle_c \tag{4.10}$$

in any order ν. Mixed cumulants with $\nu_1 \neq \nu_2$ then are zero.

Now we return to the general case in which correlated x_k are included as well. The cumulants of order one are equal to the moments of order one, that is to say, they are the mean values of x_k:

$$\langle x_k \rangle_c = \langle x_k \rangle. \tag{4.11}$$

The cumulants of order two are

$$\langle x_k x_l \rangle_c = \langle x_k x_l \rangle - \langle x_k \rangle \langle x_l \rangle. \tag{4.12}$$

These equations and corresponding expressions for some higher orders can be obtained easier by expanding the generating function not of x but of the *fluctuation, variation* or *deviation*

$$\Delta x = x - \langle x \rangle \tag{4.13}$$

of x. The second cumulant eq. (4.12) can also be writen

$$\langle x_k x_l \rangle_c = \langle \Delta x_k \Delta x_l \rangle. \tag{4.14}$$

It is called the *correlation function* of the fluctuations Δx_k, Δx_l if $k \neq l$, and the *variance* of x_k if $k = l$:

$$\langle x_k^2 \rangle_c = \langle (\Delta x_k)^2 \rangle. \tag{4.15}$$

The variance gives a measure of the dispersion of x_k around the mean value $\langle x_k \rangle$. We can interpret

$$Q_{kl} = \langle \Delta x_k \, \Delta x_l \rangle \tag{4.16}$$

to be the elements of a matrix Q which is called *correlation matrix* of the fluctuations.

In most of the practical cases only the cumulants of the first and second order are considered. At least, they are the most important, as is true also for the moments.

Without proof it may be mentioned that a cumulant of order ν is a polynomial of moments the highest order of which is also ν. The inversion is true as well.

1.1.5 The Normal Distribution

The particular importance of the cumulants of first and second order raises the question with which probability distribution $\rho(x)$ they are associated if all higher cumulants vanish. If nothing else is known than these low order cumulants, it is a practice to put all higher order cumulants equal to zero to obtain an approximation for $\rho(x)$ which is not burdened with an unjustified pecularity.

For convenience we consider first the one-dimensional case with only one random variable x with any real value. We can connect the *Fourier* transform of $\rho(x)$,

$$\hat{\rho}(\xi) = \int_{-\infty}^{\infty} dx\, e^{i\xi x} \rho(x) = \langle e^{i\xi x} \rangle \tag{5.1}$$

with the generating function $\Gamma(\alpha)$ of the cumulants eq. (4.7)

$$\Gamma(i\xi) = \ln \hat{\rho}(\xi) = i\xi\, C_1 - \frac{1}{2}\, C_2\, \xi^2 . \tag{5.2}$$

C_1 is the mean value $\langle x \rangle$ of x, and i is the imaginary unit. The inversion of the *Fourier* transform

$$\rho(x) = \frac{1}{2\pi} \int_{-\infty}^{\infty} d\xi\, e^{-i\xi x} \hat{\rho}(\xi) \tag{5.3}$$

thus yields

$$\rho(x) = \frac{1}{2\pi} \int_{-\infty}^{\infty} d\xi\, \exp\left(-i\xi s - \frac{1}{2} C_2\, \xi^2\right), \tag{5.4}$$

where s is the fluctuation Δx of eq. (4.13). With

$$\xi + i\, \frac{s}{C_2} = \zeta \tag{5.5}$$

we can write

$$\rho(x) = \exp\left(-\frac{s^2}{2C_2}\right) \frac{1}{2\pi} \int d\zeta\, \exp\left(-\frac{1}{2} C_2\, \zeta^2\right). \tag{5.6}$$

The integration path runs in the complex ζ-plane parallel to the real axis because the imaginary part of ζ runs from $-\infty$ to $+\infty$. It is, however, allowed to shift the integration path into the real axis of ζ without changing the integral. This is the case because the shift goes only across a stripe of the ζ-plane in which the integrand is regular, and zero at the ends of the stripe at infinity. The integral over ζ thus is a real constant and we obtain

$$\rho(x) = A \exp\left(-\frac{s^2}{2\langle s^2 \rangle}\right). \tag{5.7}$$

Here we wrote C_2 down in full as the variance of the fluctuation

$$s = x - \langle x \rangle = \Delta x \tag{5.8}$$

of x. The function $\rho(x)$ is called a *normal* or *Gaussian* distribution. The constant A is determined by the normalization condition for a probability distribution

$$\int\limits_{-\infty}^{\infty} dx\, \rho(x) = 1. \tag{5.9}$$

The integral

$$I = \int\limits_{-\infty}^{\infty} dx \, \exp\left(-x^2\right) = \pi^{-1/2} \tag{5.10}$$

occurs very frequently in statistics and can easily be obtained by evaluating the surface integral over an (x, y)-plane

$$I^2 = \int dx\, dy\, \exp\left(-x^2 - y^2\right) = 2\pi \int\limits_0^{\infty} dr\, r \exp\left(-r^2\right). \tag{5.11}$$

So we obtain

$$A = \left(2\pi \langle s^2 \rangle\right)^{-1/2}. \tag{5.12}$$

As shown in Figure 4, the function $\rho(x)$ has a bell shaped profile with the width proportional to $\langle s^2 \rangle^{1/2}$. The function is maximum at $x = \langle x \rangle$, and is often called a *Gauss bell*.

Now it is not difficult to generalize to more than one random variable x_k. Generally for any distribution $\rho(x)$ in a more-dimensional parameter space x the generating function $Z(\alpha)$ of the moments is connected with the *Fourier* transform

$$\hat{\rho}(\xi) = \int d^d x \, e^{i\xi x} \rho(x) = Z(i\xi) \tag{5.13}$$

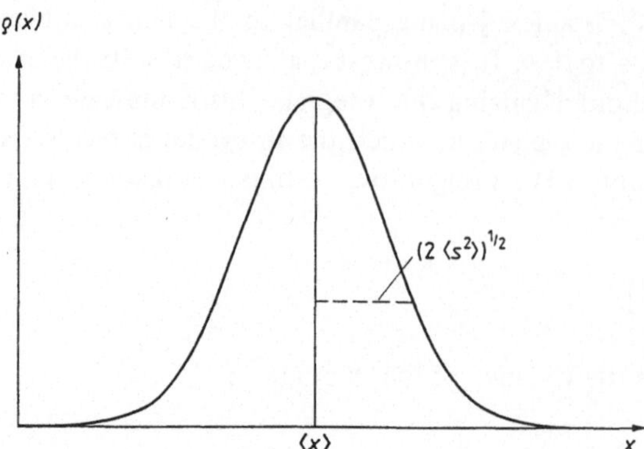

Figure 4 The normal distribution (*"Gaussian* bell")
$\rho(x) = A\,\exp(-s^2/2\langle s^2\rangle)$

of $\rho(x)$. The function $\hat{\rho}(\xi)$ is also called the *characteristic function* of the random variables x. We seek the distribution $\rho(x)$ the only non-zero cumulants of which are of the order one and two. As generalization of eq. (5.4) we obtain

$$\rho(x) = (2\pi)^{-d}\int d^d\xi\,\exp\left(-i\xi\,\Delta x - \frac{1}{2}\,\xi Q\xi\right).\tag{5.14}$$

d is the dimension of the x-space. Q is the correlation matrix of the fluctuations with the matrix elements:

$$Q_{kl} = \langle\Delta x_k\,\Delta x_l\rangle.\tag{5.15}$$

These elements are the cumulants of the second order, eqs. (4.12, 4.15). As Q is symmetric, there is an orthogonal transformation

$$\xi = Uy\tag{5.16}$$

which converts the quadratic form of Q into a diagonal one:

$$\xi Q\xi = \sum_{r=1}^{d} y_r^2\,\gamma_r = y\,\gamma y.\tag{5.17}$$

γ is a diagonal matrix with the diagonal elements γ_r. Orthogonality of U means that the transposed U^T of the matrix U is equal to the inverse U^{-1}. The determinant of U is 1. Thus

$$\rho(x) = \prod_{r=1}^{d}\frac{1}{2\pi}\int d\eta_r\,\exp\left(-i\eta_r\,\Delta y_r - \frac{1}{2}\,\eta_r^2\,\gamma_r\right).\tag{5.18}$$

This is a product of integrals of the type in eq. (5.4) equal to eq. (5.7). With a normalization constant B,

$$\rho(x) = B\,\exp\left(-\frac{1}{2}\,\Delta y\,\gamma^{-1}\Delta y\right).\tag{5.19}$$

Introducing the matix

$$g = U^{-1} \gamma^{-1} U \tag{5.20}$$

we obtain

$$\rho(x) = B \exp\left(-\frac{1}{2} sgs\right), \tag{5.21}$$

where we again use the abbreviation s for the fluctuations

$$s = \Delta x = U \Delta y. \tag{5.22}$$

Comparison of eq. (5.20) with

$$Q = U^{-1} \gamma U \tag{5.23}$$

shows that the matrix g occurring in the normal distribution $\rho(x)$ is the reciprocal of the correlation matrix of the fluctuations,

$$Q = \langle s \cdot s \rangle = g^{-1}. \tag{5.24}$$

Q has the matrix elements

$$Q_{kl} = \langle s_k s_l \rangle. \tag{5.25}$$

The correlation matrix of the fluctuations of the variables y is

$$\langle \Delta y \cdot \Delta y \rangle = U Q U^{-1} = \gamma. \tag{5.26}$$

This means, that the y_k, y_l are uncorrelated,

$$\langle \Delta y_k \Delta y_l \rangle = 0 \tag{5.27}$$

for $k \neq l$ and therefore distinguished variables. They are called *normal variables*.

At the beginning of this chapter we gave an argument for the distinction of the normal distribution. In the next chapter we shall give still another argument which shows why this distribution is expected to be an adequate standard approximation in many applications.

*1.1.6   The Central Limit Theorem

We shall show that the sum of independent random variables of the same type which become very large in number tends to assume a normal distribution. Thus the normal distribution is distinguished in statistics once more. To give an illustration of the situation which will be discussed below, we consider the so-called *random walk*. The random walk can be used as a model for the motion of a *Brownian particle*. This is a particle large enough to be visible in a light microscope, but so small that the irregular bombardment of the molecules of a surrounding liquid gives rise to an unceasing and irregular tumbling motion of the particle. We assume now that we register the position of such a particle only at discrete times

$$t = n\tau \qquad (n = 0, 1, 2, \dots), \tag{6.1}$$

separated by a finite time interval τ in which any influence of the last step to the next one is totally destroyed. We are interested in the motion in one direction only. The step in this direction between the time $(n-1)\tau$ and the time $n\tau$ will be called l_n. The random quantities l_n are uncorrelated with one another, and if there is no external force acting on the particle, the mean value of l_n is zero. The mean value of the whole path

$$x = \sum_{n=1}^{N} l_n \tag{6.2}$$

after the time $N\tau$ will then be zero, too; but $\langle x^2 \rangle$ will not.

Having given this illustration, we shall formulate the problem generally. Given a set of random quantities l_n which are uncorrelated and vanish in the average:

$$\langle l_n \rangle = 0. \tag{6.3}$$

The variance of each one is

$$\langle l_n^2 \rangle = l^2. \tag{6.4}$$

The variance of x in eq. (6.2) thus is

$$\langle x^2 \rangle = Nl^2. \tag{6.5}$$

We introduce the quantity

$$y = N^{-1/2} l^{-1} x \tag{6.6}$$

which has the variance 1. As all l_n are uncorrelated with one another, their mixed cumulants of any order vanish. The cumulants of order ν

$$\langle l_n^\nu \rangle_c = \lambda_\nu \tag{6.7}$$

are equal for all n and shall remain finite in any order The cumulants of y are

$$\langle y^\nu \rangle_c = N^{-\frac{\nu}{2}} l^{-\nu} \sum_{n=1}^{N} \langle l^\nu \rangle_c = N^{1-\frac{\nu}{2}} l^{-\nu} \lambda_\nu. \tag{6.8}$$

We now consider the case that the number N of the random variables l_n becomes arbitrarily large. In the limit of infinite N all cumulants of y vanish with exception of the second one. The characteristic function is

$$\hat{\rho}(\xi) = \exp\left(-\frac{1}{2}\xi^2\right). \tag{6.9}$$

The corresponding probability density of y is the normal distribution

$$\rho(y) = \frac{1}{2\pi} \exp\left(-\frac{1}{2}y^2\right). \tag{6.10}$$

This result is called the *central limit theorem* of the probability theory. In many cases and under very general conditions the theorem gives reason to try the assumption of a normal distribution as a useful appproximation. Deviations from the normal distribution correspond to cumulants of y of higher order than two and are an indication of correlations between the elementary random quantities l_n.

1.2   Information Measures

During World War II, in the USA the problem of decoding enciphered messages gave rise to the development of a new branch in the theory of communication, the "information theory". This development is connected in particular with the names *C. Shannon* and *N. Wiener*. There new concepts, the "information measures", are of fundamental importance. Such measures are numbers attached to probability distributions and thus functions of the probabilities if the distributions belong to enumerable events. If the distributions are given by probability densities belonging to continuous events, the measures are functionals of these densities. A measure of this kind, distinguished in different respect, is the *Shannon information*. Having arisen in communication theory it turned out to be directly connected with entropy in statistical mechanics, a fundamental concept in physics which was created more than half a century earlier. This connection gave rise to many discussions about the interpretation of entropy. In effect all these discussions result in the question about the role of probability in thermodynamics.

In this section we shall be concerned mainly with two information measures. The first one is the *Shannon* information mentioned above. This measure depends on only one probability distribution and is closely related to thermodynamic entropy. The other one is the so-called *Kullback information* or *information gain*, which depends on two probability distributions and is closely related to the *availability* or *exergy*, another thermodynamic quantity of fundamental importance. Under certain conditions, this quantity is proportional to *produced entropy* in the interior of a physical system.

Several features of these measures in general statistics can be found again in the basic structure of macroscopic phenomenological thermodynamics, where they had been well known long before the information theory was developed.

1.2.1   *Shannon* Information

We shall first consider a sample set of n events U_i and seek a measure which makes it possible to compare different probability distributions over this set. The whole set of the probabilities P_i of a distribution shall be abbreviated by P. The desired measure should make it possible to decide which one of two distributions gives more knowledge about the question which event of the sample set will occur. For instance, a distribution with only one P_j equal to one while all other P_i vanish, i.e.

$$P_i = \delta_{ij}, \tag{1.1}$$

gives the best possible knowledge, the *certainty*. In contrast with this, the *homogeneous* distribution in which all events U_i have equal probability

$$P_i = \frac{1}{n} \tag{1.2}$$

does not prefer any U_i and belongs to minimum knowledge.

Bit-Numbers. One way to construct such a measure starts with the question how long a message has to be to inform an observer about the fact that the event U_i has indeed occurred if he only knows the probability P_i. Of course we ask for the shortest possible message avoiding any redundance. The length of this message is a measure of the knowledge the observer is missing for the prediction whether U_i will occur or not. We require that the measure depends on P_i only and not on any other individual feature of the events of the sample set. We call this measure $b(P_i)$.

In physics the results of a measurement represent a sample set, say $U_1, U_2, \dots$. If with each U_i in principle all measurable quantities of a physical system are given, we say that U_i fixes the state of the system. Let us call U_i simply the "state" of the system. The term "state", however, will be also used in a more extended way. Therefore we call these states U_i *pure states*. Also, if the pure state of the system is not known exactly but only the probability distribution P_i over the sample set of all U_i, we speak of a "state", or better a *mixture state* to distinguish it from a pure state.

Now we start with a fundamental assumption. If U_i and U'_α are two different sample sets having no element in common, we can form the sample set of all "compound" events $U''_{i\alpha}$, the products $U_i U'_\alpha$. In the case that U_i are pure states of one physical system and U'_α those of another one, we can interpret these two systems to be parts of a larger system which is composed of these two "subsystems". The pure states of the composed system are $U''_{i\alpha}$. The two subsystems are stochastically independent of each other if the two sample sets are never correlated with each other. This means that always

$$P''(U''_{i\alpha}) = P(U_i) P'(U'_\alpha). \tag{1.3}$$

Let us designate the logarithm of P with η, and $b(P)$ with $f(\eta)$. We distinguish the quantities belonging to the three systems by the use of dashes as already done. For the two independent systems we have to require that the length of the messages is additive:

$$f''(\eta + \eta') = f(\eta) + f'(\eta'). \tag{1.4}$$

This is a functional equation yielding that

$$\frac{\partial f}{\partial \eta} = \frac{\partial f'}{\partial \eta'}. \tag{1.5}$$

This is possible only if both sides of the equation are constant. This constant which may be called $-\kappa$ is the same for all systems because we can apply the result to the compositions of two arbitrary systems. We obtain

$$b(P) = -\kappa \ln P + C. \tag{1.6}$$

The constant κ has to be positive to make b positive. b has to be zero for a "certain" event. Therefore C has to be zero.

A sample set of only two events is called a *simple alternative* or a *bit* (binary digit). If the two events have the same probability $1/2$, the measure $b(P)$ is

$$b\left(\frac{1}{2}\right) = \kappa \ln 2. \tag{1.7}$$

Let us now consider a special sample set which has the following particular property. It can be divided into two separate subsets of equal probability $1/2$. Each one of these shall have this particular property again, as shall their two subsets and all other subsets which arise from subsequent divisions of this kind. (This particular property will always be realized in good approximation if the number n of the elements of the sample set is large enough. Deviations from this property will occur only in the latest steps which do not matter here. In a common family game such successive division into always two subsets is the best way to find out a certain object by as few yes-no questions as possible.) For such a "fair" sample set we can give the shortest message with the length eq. (1.6) in particular by the sequence of bits belonging to the described successive divisions. The number of these bits is

$$B_i = \frac{b(P_i)}{b(\frac{1}{2})} = -\frac{\ln P_i}{\ln 2} \, . \tag{1.8}$$

For convenience

$$b(P_i) = -\ln P_i \tag{1.9}$$

is usually called *bit-number*. This means that $\ln 2$ is chosen as the unit of one bit (sometimes called one *bin*). In other words, κ is chosen to be one. The definition of bit-number by eq. (1.9) be extended to any sample set, even if it is not of the described particular "fair" type. Then B_i is allowed not to be an integer.

Information of a Distribution. If such messages are given very often, the message belonging to the particular result U_i will occur with the relative frequency P_i. Per event we need in the average the bit-number

$$\langle b \rangle = - \sum_{i=1}^{n} P_i \ln P_i \, . \tag{1.10}$$

The larger this number, the less knowledge is given by the distribution. Therefore its negative value is defined as the information measure of the distribution P, called the *Shannon information*[2])

$$I(P) = \sum_{i=1}^{n} P_i \ln P_i \, . \tag{1.11}$$

This measure is by definition always negative. For the distribution eq. (1.1), called *certainty*, it becomes zero, which indeed is the maximum of $I(P)$. The minimum of $I(P)$ is found by solving the equation for infinitesimal changes δP_i of P_i

$$\delta I = \sum_{i} (1 + \ln P_i) \, \delta P_i = 0 \tag{1.12}$$

under the restrictive condition of the normalization eq. (3.2) of chapter 1.1.3 which means

$$\sum_{i} \delta P_i = 0. \tag{1.13}$$

---

[2]) *C. Shannon* and *W. Weaver*, The Mathematical Theory of Communication (University Press of Illinois, Urbana 1949).

With a *Lagrangian* multiplier λ this yields

$$\sum_i (\ln P_i - 1 - \lambda)\, \delta P_i = 0. \tag{1.14}$$

Due to eq. (1.13) only $n - 1$ variations δP_i are independent from one another. We can choose λ to make one of the brackets in eq. (1.14) equal to zero. The remaining $n - 1$ brackets must vanish being the coefficients of the independent δP_i. Thus all brackets are zero and all P_i are equal. This is the homogeneous distribution eq. (1.2). – We see that the requirements are fulfilled and that $I(P)$ goes to the minimum for eq. (1.2). If we compare different sample sets, the information measure of the homogeneous distribution eq. (1.2)

$$I(P) = -\ln n \tag{1.15}$$

decreases with increasing number n of events. Indeed the more events are possible, the less we know about which event will occur.

$I(P)$ is a symmetric function of all P_i, dependent on no other individual features of the events but the probabilities.

Continuous events. The case that the events of the sample set are continuous, i.e. that their mapping requires a continuous parameter space x, needs particular consideration. The probability distribution then is described by a probability density $\rho(x)$. We can refer to the case of discrete events by dividing the parameter space into very small cells of equal d-dimensional volume $\Delta^d x$ situated at discrete lattice points $x^{(i)}$. If $\rho(x)$ is changing in x-space smoothly enough and if $\Delta^d x$ is small enough, the probability of the cell i is

$$P_i = \rho(x^{(i)})\, \Delta^d x. \tag{1.16}$$

Thus

$$I(P) = \sum_i \Delta^d x\, \rho(x^{(i)}) \ln \rho(x^{(i)}) + \ln \Delta^d x. \tag{1.17}$$

We are interested in the comparison of different distributions $\rho(x)$ in the same parameter space x. This is done best by fixing the size $\Delta^d x$ of the cells. Then we may drop the last term in eq. (1.17) and define an information measure of each density by the remaining part. In the limit of vanishing $\Delta^d x$ it is

$$I(\rho) = \int d^d x\, \rho \ln \rho. \tag{1.18}$$

This measure is different from $I(P)$ but remains finite in the limit whereas $I(P)$ diverges. We indicate this distinction by the argument ρ instead of P.

Some words should be said about the particular use of the term "information" in this context. It should be emphasized that the *Shannon* information is a measure of the knowledge about a very special question only, i.e. about the question which event of a sample set will occur and of nothing else. Ignoring this restriction can give rise to confusion in discussions about the role of information in science. To give an example, we regard the homogeneous distribution eq. (1.2) in which each event has the same probability. This distribution belongs to minimum information. It can, however, be obtained in very different ways. This distribution can for instance be assumed because of the lack of any prior observation in order to avoid any unjustified preference of some events. It can, in other cases, be the result of detailed and elaborate prior observation. (When more than three thousend years ago in old China it was found that the probability of the birth of a boy was nearly equal to the birth of a girl, this finding was based on extended experience. By the way, observations in Europe made later, yielded slight deviations.) In both cases the knowledge of the observer is very different, but not the special knowledge expressed by $I(P)$. The observer in both cases is equally unable to predict which event will occur. Therefore the *Shannon* information is the same. For this measure it makes no difference how the probability distribution was gained.

1.2.2   Information Gain

In this chapter we shall be concerned with another measure which is not an alternative to the *Shannon* information. The latter depends on only one probability distribution. The measure we shall now discuss depends on two distributions. It is a measure of the excess information contained in a distribution P in comparison to another "reference" distribution P'. It is closely related to the *availability* or *exergy* already mentioned in the introductory part of this subsection. Availability is a further fundamental quantity in thermodynamics different from entropy and will be discussed in chapter 2.3.6.

If P and P' are probability distributions over the same sample set of events U_i, then

$$b(P_i') - b(P_i) = \ln \frac{P_i}{P_i'} \tag{2.1}$$

is the bit-number necessary to change the probability P_i' into P_i by a message. This message may for instance be based on the results of a new measurement. It yields a correction of the probability P_i' into P_i. The mean value of the bit-number eq. (2.1) formed with the weights of the "corrected" distribution P is

$$K(P, P') = \sum_{i=1}^{n} P_i \ln \frac{P_i}{P_i'} \tag{2.2}$$

and is called *information gain* or *Kullback information*. It was introduced by *S. Kullback* in 1951[3]. It should be stressed that this measure is asymmetrical in P and P', which is an essential property of the corresponding quantities in thermodynamics as well. Whereas the bit-number difference eq. (2.1) can assume both signs, the measure K is never negative. This is a consequence of the inequality

$$\ln x \geqslant 1 - \frac{1}{x}, \tag{2.3}$$

valid for all $x > 0$, with the normalization eq. (3.3) of chapter 1.1.2.

---

[3] *S. Kullback*, Ann. Math. Statistics **22**, 79 (1951);
   S. Kullback, Information Theory and Statistics (Wiley, New York 1951).

To make K finite and unique, we require that no vanishing P_i' occurs. This will tacitly be supposed whenever we apply the measure K. This means that we shall never take into account totally improbable events U_i in the distribution.

K vanishes if and only if the distributions P and P' are exactly identical. Then it is minimum. We can see this by seeking minimum K. The corresponding equation

$$\delta K = \sum_i \left[\ln \frac{P_i}{P_i'} + 1 \right] \delta P_i = 0 \tag{2.4}$$

for infinitesimal δP_i with the restriction

$$\sum_i \delta P_i = 0 \tag{2.5}$$

yields by use of the *Lagrangian* multiplier $\lambda - 1$

$$\ln \frac{P_i}{P_i'} = \lambda \tag{2.6}$$

for all i. Thus P_i is proportional to P_i'. Due to the normalization, this is only possible if λ vanishes, i.e. if P and P' are identical.

The *Shannon* information can be regained from the *Kullback* information in the special case that the reference distribution is the homogeneous one. This yields

$$K = I(P) + \ln n. \tag{2.7}$$

Thus all properties of the *Shannon* information can be deduced from the *Kullback* information as the more general concept.

$K(P, P')$ is a convex function of P because

$$\frac{\partial^2 K}{\partial P_i \, \partial P_j} = \frac{1}{P_i} \, \delta_{ij} \geqslant 0. \tag{2.8}$$

Therefore $I(P)$ is a convex function of P as well.

Continuous Events. The transition from discrete to continuous events is simpler for K than for I because $\ln \Delta^d x$, which occurs in eq. (1.17), is cancelled in K. With eq. (1.16) we obtain

$$K(P, P') = \sum_i \Delta^d x \, \rho(x_i) \ln \frac{\rho(x_i)}{\rho'(x_i)} . \tag{2.9}$$

In the limit of vanishing $\Delta^d x$, this expression goes over directly into

$$K(\rho, \rho') = \int d^d x \, \rho \ln \frac{\rho}{\rho'} . \tag{2.10}$$

We see that no change of the definition is necessary for K, whereas it was needed for the transition from $I(P)$ to $I(\rho)$. The absence of $\ln \Delta^d x$ in K leads to another distinguishing property of the *Kullback* information. Any one-to-one transformation of the coordinates x of the parameter space, say into $\tilde{x}$, changes the density $\rho(x)$ into

$$\tilde{\rho}(\tilde{x}) = \rho(x) \, \mathrm{Det} \, \frac{\partial x}{\partial \tilde{x}}, \tag{2.11}$$

where the occurring factor is the *Jacobian* determinant of the transformation. ρ' is transformed with the same factor Therefore $I(\rho)$ and $I(\tilde{\rho})$ are generally different. The *Jacobian*, however, is cancelled in

$$K(\tilde{\rho}, \tilde{\rho}') = K(\rho, \rho'). \tag{2.12}$$

K is invariant with respect to the transformation.

*1.2.3 Stochastic Matrices

We shall discuss the behavior of the information gain $K(P, P')$ with respect to a special class of transformations of the distributions P, P'. This will show a further distinguishing property of the measure K. It will, moreover, yield relations which will be helpful for the deduction of results we shall need elsewhere in this book.

To begin with, we assume that the transition of a physical system from a pure state U_i to another U_j during a certain time τ occurs with the *transition probability*

$$\Lambda_{ji} = P(U_j \mid U_i). \tag{3.1}$$

This is the conditional probability that the system is in state U_j after time τ if it was in U_i before. If it, however, started in the mixture state P given by the distribution P, it will after the time τ be in a mixture state $\hat{P}$ given by

$$\hat{P}_j = \sum_i \Lambda_{ji} P_i. \tag{3.2}$$

The special distribution called *certainty*

$$P_i - \delta_{ij} \tag{3.3}$$

was already introduced by eq. (1.1), and yields

$$\Lambda_{ji} \geqslant 0, \tag{3.4}$$

$$\sum_{j=1}^{n} \Lambda_{ji} = 1, \tag{3.5}$$

as $\hat{P}$ is a probability distribution as well. These two relations definie Λ_{ji} as a co-called *stochastic matrix*.

Under very general conditions, the transition probability Λ_{ji} will be independent of P. Then the transformation of P into $\hat{P}$ is linear. We also find such linear transformations of a distribution P into another one $\hat{P}$ over the same sample set of events in other connections. Therefore we shall discuss them more generally.

If two distributions P, P' over the same sample set are linearly transformed by the same stochastic matrix Λ_{ji},

$$\hat{P}_j = \sum_i \Lambda_{ji} P_i , \tag{3.6}$$

$$\hat{P}'_j = \sum_i \Lambda_{ji} P'_i , \tag{3.7}$$

the information gain never increases:

$$K(\hat{P}, \hat{P}') \leqslant K(P, P'). \tag{3.8}$$

That is to say that the transformation can destroy excess knowledge yet never can produce it. This can be proved again by the use of inequality (2.3), starting with

$$K(P, P') - K(\hat{P}, \hat{P}') = \sum_{ij} \Lambda_{ji} P_i \ln \frac{P_i \hat{P}'_j}{P'_i \hat{P}_j} . \tag{3.9}$$

The right hand side is never less than

$$\sum_{ij} \Lambda_{ji} P_i \left(1 - \frac{P'_i \hat{P}_j}{P_i \hat{P}'_j} \right) = 0. \tag{3.10}$$

This theorem eq. (3.8) becomes important in particular if P' is an invariant distribution P^0 of the transformation satisfying

$$\sum_i \Lambda_{ji} P_i^0 = P_j^0 . \tag{3.11}$$

Then

$$K(\hat{P}, P^0) \leqslant K(P, P^0). \tag{3.12}$$

A special stochastic matrix is one which leaves the homogenenous distribution invariant. It does never change minimal knowledge which we may call *ignorance*. Such a matrix satisfies not only eq. (3.5) but moreover the equations

$$\sum_{i=1}^{n} \Lambda_{ji} = 1, \tag{3.13}$$

which differ from eq. (3.5) by the exchanged summation indices. It is called a *bistochastic matrix*. For a transformation by such a matrix holds

$$I(\hat{P}) \leqslant I(P). \tag{3.14}$$

As a rule, knowledge is distroyed by such a transformation.

Remark: Equations of the type of eq. (3.2) are often used to describe the development of an ensemble of independent equal systems in different states U_i by the change from P to $\hat{P}$. Then P_i is interpreted as the relative frequency with which the systems are found in state U_i. In this case the transformation matrix Λ of the transition probabilities is independent of these relative frequencies and eq. (3.2) is linear in P. An example where Λ is dependent on the probability distribution P is the *Boltzmann* equation in which the events U_i are dynamical states of one particle in a gas. There the transition probabilities from one state U_i into another one depend on the state of other particles in the gas and thus on the probability of these states. Then eq. (3.2) is not any more linear in P. Linearity of this equation is decisively dependent on which "level" we describe a physical system; i.e., which sample set of events U_i we are speaking of. In the *Boltzmann* equation the events U_i are one-particle states and not microstates of the whole gas. They are rather *mesoscopic* states, as many microstates belong to each such state.

1.3   Generalized Canonical Distributions

In many practical cases a probability distribution is not gained by direct observation of relative frequencies but by the mean values of certain quantities. This is a standard situation in statistical thermodynamics. A macroscopic *thermal state* can be described by variables which in statistical mechanics are interpreted as mean values over many microstates. The question is, how can we draw conclusions about a probability distribution from the knowledge of some mean values? In general this is an open question. There is, however, an answer in the special case that the sample set of events to which the distribution belongs has a particular property which is called *a priori equal probability*. This concept will be explained in the next chapter. The method to find the adequate probability distribution is called the *unbiased guess* and is based on the use of the *Shannon* information. As will be pointed out in the next main section, the mathematical structure of this method is identical with a construction of probability distributions over microstates in thermodynamics by requiring maximum entropy under certain conditions. This construction was already developed in the last century, long before an information measure was known. It gave a foundation of the so-called *canonical distribution* and of some of its generalizations. The unbiased guess is not restricted to thermodynamics and forms a basis of these distributions in the framework of general statistics not at all restricted to physics. It is based on an interpretation given by *E. T. Jaynes* in 1957 which casts new light on the connection between probability and experience. The method and its result which we shall call *generalized canonical distribution* will be explained in the next chapter.

In chapters following thereafter we shall be concerned with relations which are implied by the structure of these generalized canonical distributions.

1.3.1   The Unbiased Guess

We shall discuss the above mentioned situation that the probability distribution P over a given sample set U is not known in detail. What is known is but a set of some mean values

$$\langle M^\nu \rangle = \sum_{i=1}^{n} P_i M_i^\nu. \tag{1.1}$$

The question is how to find an adequate distribution P corresponding to this knowledge. Of course the number of these mean values is supposed to be smaller than the number n

of events in the sample set. In the cases we are interested in, n is extremely large compared with the former number. Then there are many distributions P which satisfy eq. (1.1). Is there an argument to distinguish one of them to be *adequate*? We can find a distinguished one if the sample set U has the particular property that all U_i have *a priori equal probability*. This property is defined in the following way.

> If no $\langle M^\nu \rangle$ is known or if no other observation has been made, then it is adequate to assume the homogenous distribution.

That is to say, in this case all events U_i have the same probability $1/n$. It does not matter on which particular reason this assumption is based in practical cases. The reasons may be very different. It may be that earlier observations about the behaviour of a physical system provide an argument. In other cases the argument may be motivated by basic physical laws. There is also a methodical argument, called the *indifference principle*.

> If nothing else is known about the events U_i there is no reason to favour one or some of them and therefore we have to suppose equal probability.

Uncritical use of this principle, however, can lead to paradox results, e.g. if no distinction is made between compound and elementary events. The assumption was tacidly made with respect to the so-called "possible" events when formulating the classical definition of probability. These had to be a priori equally probable "elementary" events of a sample set out of which the compound events of consideration were built up.

> The unbiased guess of *Jaynes*[4] is the method of seeking this one of all possible distributions P which comprises minimum information $I(P)$. Any other would comprise unjustified prejudices.

This means, we have to seek P satisfying eq. (1.1) with given values $\langle M^\nu \rangle$ as well as

$$\sum_{i=1}^{n} P_i = 1, \tag{1.2}$$

$$I(P) = \sum_{i=1}^{n} P_i \ln P_i = \min. \tag{1.3}$$

For any infinitesimal change δP of P therefore it holds that

$$\sum_i M_i^\nu \, \delta P_i = 0, \tag{1.4}$$

$$\sum_i \delta P_i = 0, \tag{1.5}$$

$$\sum_i (1 + \ln P_i) \delta P_i = 0. \tag{1.6}$$

---

[4] *E. T. Jaynes*, Phys. Rev. **106**, 620 (1957).

We can multiply each eq. (1.4) by a *Lagrange* factor λ_ν, eq. (1.5) by another one, say $-(\Psi + 1)$ and obtain by summation over these equations

$$\sum_i (\ln P_i - \Psi + \lambda_\nu M_i^\nu)\, \delta P_i = 0. \tag{1.7}$$

Here and in all that follows, repeated Greek super- and subscripts in a product indicate a summation over all ν. We proceed similarly to eq. (1.14) of chapther 1.2.1. In eq. (1.7) the same number of brackets as *Lagrange* multipliers Ψ, λ_ν occur can be made equal to zero by the adequate choice of these multipliers. The remaining δP_i are not restricted by the conditions of eqs. (1.4, 1.5) the number of which is equal to the number of *Lagrange* multipliers as well. Therefore, all brackets vanish for the adequate values λ_ν, Ψ:

$$P_i = \exp(\Psi - \lambda_\nu M_i^\nu). \tag{1.8}$$

We call this result a *generalized canonical distribution*.

In the case of continuous events x we habe to seek the distribution $\rho(x)$ for which

$$I(\rho) = \int d^d x\, \rho(x) \ln \rho(x) \tag{1.9}$$

is minimum under the conditions

$$\int d^d x\, \rho(x) = 1, \tag{1.10}$$

$$\int d^d x\, \rho(x)\, M^\nu(x) = \langle M^\nu \rangle. \tag{1.11}$$

For a variation $\delta \rho(x)$ of $\rho(x)$ then

$$\int d^d x\, (1 + \ln \rho)\, \delta\rho = 0, \tag{1.12}$$

$$\int d^d x\, \delta\rho = 0, \tag{1.13}$$

$$\int d^d x\, M^\nu \delta\rho = 0, \tag{1.14}$$

which yields with the same *Lagrange* multipliers as before

$$\int d^d x\, (\ln \rho - \Psi + \lambda_\nu M^\nu)\, \delta\rho = 0. \tag{1.15}$$

The choice of the function $\delta\rho(x)$ is not totally free because eqs. (1.13, 1.14) are restrictive. To begin with, the *Lagrange* mutlipliers are free. We can satisfy eq. (1.15) by putting the bracket in it equal to zero and we obtain

$$\rho(x) = \exp(\Psi - \lambda_\nu M^\nu(x)). \tag{1.16}$$

As the number of the *Lagrange* multipliers is equal to the number of the conditions, we can fulfil the conditions by an adequate choice of these parameters. The probability density of eq. (1.16) is also called a "generalized canonical distribution".

The described connection of the "premise", given by the mean values $\langle M^\nu \rangle$ with the "consequence", the generalized canonical distribution as the result of the unbiased guess, represents a typical conclusion of an "inductive logic". So is called a conclusion which is not a stringent syllogism in the sense of a pure analysis of the premise but incorporates other empirical or methodical elements.

1.3.2   Properties of the Generalized Canonical Distribution

In this chapter relations will be obtained which result from the particular structure of the generalized canonical distribution. Without being restricted to physics, they are fundamental in thermodynamics and determine the specific structure of this field in physics. It is useful to bring these relations into connection with the concept of the *Legendre* transformation to give them more uniformity and transparency. Therefore we shall first introduce this concept.

The *Legendre* transformation. To explain this transformation, we first consider the example of a point moving along a line. The length of the path it has passed after the time t may be $\Psi(t)$. The velocity is

$$M = \frac{d\Psi}{dt}. \tag{2.1}$$

If we know only Ψ as function of M, we are not able to reconstruct the whole motion $\Psi(t)$ (if we register only milage and demarcation on the speedometer without time, we cannot say when the trip began). The function

$$I(M) = \Psi(t) - Mt \tag{2.2}$$

of M, however, allows to reconstruct $\Psi(t)$ correctly. First we obtain t by

$$\frac{dI}{dM} = \frac{d\Psi}{dt}\frac{dt}{dM} - M\frac{dt}{dM} - t, \tag{2.3}$$

which due to eq. (2.1) is

$$\frac{dI}{dM} = -t. \tag{2.4}$$

As with this finally M is determined as a function of t, eq. (2.2) yields $\Psi(t)$. In Figure 5 a curve $\Psi(t)$ is plotted and the construction of I is shown for one value t. We obtain I as the section of the Ψ-axis cut off by the tangent of Ψ in point t. Therefore the connection eqs. (2.1, 2.2) between $\Psi(t)$ and $I(M)$ is called a *contact transformation*, as synonym of *Legendre transformation*.

This transformation can be generalized to more than one independent variable. With a view to the designations already used in the preceding chapter, we replace t with a set of variables λ_ν. A function $\Psi(\lambda)$, shall be given. If

$$M^\nu = \frac{\partial\Psi}{\partial\lambda_\nu}, \tag{2.5}$$

the *Legendre* transform of Ψ is

$$I(M) = \Psi(\lambda) - \lambda_\nu M^\nu. \tag{2.6}$$

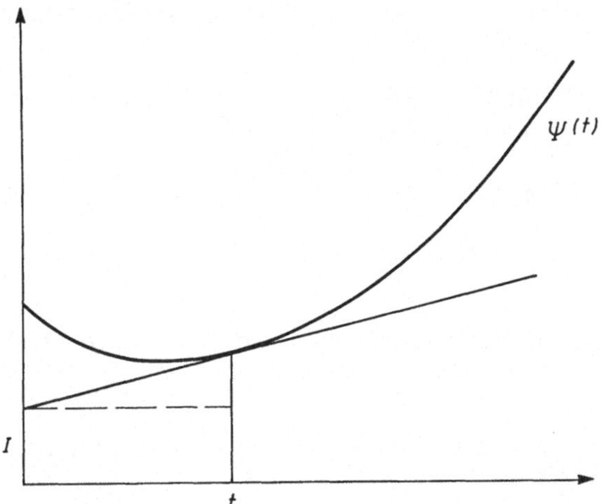

Figure 5 The connection between $\Psi(t)$ and $I(M)$ in the "contact" or *"Legendre* transformation"

Indeed, this function of the set M of all M^ν allows to reconstruct $\Psi(\lambda)$. It holds that

$$\frac{\partial I}{\partial M^\nu} = \frac{\partial \Psi}{\partial \lambda_\sigma} \frac{\partial \lambda_\sigma}{\partial M^\nu} - M^\sigma \frac{\partial \lambda_\sigma}{\partial M^\nu} - \lambda_\nu, \tag{2.7}$$

which, due to eq. (2.5), is

$$\frac{\partial I}{\partial M^\nu} = - \lambda_\nu. \tag{2.8}$$

As with this finally $M(\lambda)$ is determined, eq. (2.6) yields $\Psi(\lambda)$.

Gibbs **fundamental equation.** We return to the generalized canonical distribution

$$P_i = \exp(\Psi - \lambda_\nu M_i^\nu) \tag{2.9}$$

or the corresponding $\rho(x)$ in the case of a continuous sample set. This discrimination does not matter. The quantity Ψ is dependent on the set λ of the parameters λ_ν because the normalization condition of the distribution requires

$$\exp(-\Psi) = \sum_i \exp(-\lambda_\nu M_i^\nu). \tag{2.10}$$

We can use the set λ_ν to characterize the distribution P instead of the mean values $\langle M^\nu \rangle$, which we shall designate in the following mostly by the shorter form M^ν.

If we change these M^ν into $M^\nu + \delta M^\nu$, the parameters λ_ν will be changed into $\lambda_\nu + \delta \lambda_\nu$. Correspondingly, the distribution P will be changed into another one, called $P + \delta P$. The information gain associated with this change is

$$K(P + \delta P, P) = \delta \Psi - \delta \lambda_\nu (M^\nu + \delta M^\nu) \geqslant 0. \tag{2.11}$$

Up to now the $\delta\lambda_\nu$ were allowed to be finite. For sufficiently small variations $\delta\lambda_\nu$, however, we can write

$$\left(\frac{\partial\Psi}{\partial\lambda_\nu} - M^\nu\right)\delta\lambda_\nu + \left(\frac{1}{2}\frac{\partial^2\Psi}{\partial\lambda_\nu\,\partial\lambda_\mu} - \frac{\partial M^\nu}{\partial\lambda_\mu}\right)\delta\lambda_\nu\,\delta\lambda_\mu \geqslant 0. \tag{2.12}$$

This inequality in particular yields

$$\frac{\partial\Psi}{\partial\lambda_\nu} = M^\nu, \tag{2.13}$$

as $\delta\lambda_\nu$ can assume both signs. The *Shannon* information of P due to eq. (1.3) is

$$I = \Psi - \lambda_\nu\,M^\nu. \tag{2.14}$$

The eqs. (2.13, 2.14) are now identical with eqs. (2.5, 2.6) defining the *Legendre* transformation from $\Psi(\lambda)$ to $I(M)$. Therefore

$$\frac{\partial I}{\partial M^\nu} = -\lambda_\nu. \tag{2.15}$$

These equations for all ν can be written in the form of only one equation

$$dI = -\lambda_\nu\,dM^\nu. \tag{2.16}$$

In thermodynamics, where it is of central importance, this equation is called *Gibbs fundamental equation*. We shall carry this name over into general statistics.

Susceptibilities. We return to the inequality (2.12). The term linear in $\delta\lambda_\nu$ vanishes yielding eq. (2.13). The quadratic term is never negative and yields by use of eq. (2.13):

$$\frac{\partial M^\nu}{\partial\lambda_\sigma}\,\delta\lambda_\sigma\,\delta\lambda_\nu = \delta\lambda_\nu\,\delta M^\nu \leqslant 0. \tag{2.17}$$

This inequality in turn can be written in the form

$$\frac{\partial\lambda_\sigma}{\partial M^\nu}\,\delta M^\nu\,\delta M^\sigma \leqslant 0. \tag{2.18}$$

We call the derivatives

$$\eta^{\nu\sigma} = \frac{\partial M^\nu}{\partial\lambda_\sigma}, \qquad \tilde{\eta}_{\sigma\nu} = \frac{\partial\lambda_\sigma}{\partial M^\nu} \tag{2.19}$$

susceptibilities because in thermodynamics they describe the sensitivity of a macroscopic quantity with respect to the change of another one. It should be stressed that $\eta^{\nu\sigma}$ and $\tilde{\eta}_{\sigma\nu}$ in general are not reciprocal to each other. $\eta^{\nu\sigma}$ is a derivative taken if all λ_τ with τ different from ν are kept constant. In $\tilde{\eta}_{\sigma\nu}$ the M^ν are kept constant. The susceptibilities can be interpreted as elements of the so-called *susceptibility matrices* η and $\tilde{\eta}$. These matrices as a whole, however, are reciprocal to each other due to the equations

$$\delta M = \eta\,\delta\lambda, \qquad \delta\lambda = \tilde{\eta}\,\delta M. \tag{2.20}$$

They have non-positive quadratic forms eqs. (2.17, 2.18). In particular, their diagonal elements are non-positive:

$$\frac{\partial M^{\nu}}{\partial \lambda_{\nu}} \leqslant 0, \qquad \frac{\partial \lambda_{\nu}}{\partial M^{\nu}} \leqslant 0. \tag{2.21}$$

As on account of eqs. (2.5, 2.8) the equations hold

$$\eta^{\nu\sigma} = \frac{\partial^2 \Psi}{\partial \lambda_{\nu} \partial \lambda_{\sigma}}, \qquad \tilde{\eta}_{\nu\sigma} = -\frac{\partial^2 I}{\partial M^{\nu} \partial M^{\sigma}}, \tag{2.22}$$

the inequalities (2.17, 2.18) also read that $I(M)$ and $-\Psi(\lambda)$ are convex functions. One has to be careful. This is true only if these quantities are expressed as functions of the correct independent variables M or λ respectively.

The susceptibility matrix η is due to eq. (2.22) symmetric. This can be seen also by the fact that it is connected with the correlation matrix

$$Q^{\nu\sigma} = \langle \Delta M^{\nu} \Delta M^{\sigma} \rangle \tag{2.23}$$

of the fluctuations of the random quantities M^{ν} as defined in chapter 1.1.4 eq. (4.16). This can be shown in the following way. The generating function of the cumulants of the variables M^{ν} is

$$\Gamma(\alpha) = \ln \langle \exp(\alpha_{\nu} M^{\nu}) \rangle. \tag{2.24}$$

Due to eq. (2.10), we may write

$$\Gamma(\alpha) = \Psi(\lambda) - \Psi(\lambda - \alpha). \tag{2.25}$$

Therefore the second cumulants are

$$Q^{\nu\sigma} = \left(\frac{\partial^2 \Gamma}{\partial \alpha_{\nu} \partial \alpha_{\sigma}} \right)_{\alpha=0} = -\frac{\partial^2 \Psi}{\partial \lambda_{\nu} \partial \lambda_{\sigma}}. \tag{2.26}$$

This means that

$$Q^{\nu\sigma} = -\frac{\partial M^{\nu}}{\partial \lambda_{\sigma}}. \tag{2.27}$$

> The susceptibility matrix η is symmetric and is the negative of the correlation matrix Q.

1.3.3 Addition of Knowledge

The method of unbiased guess will now be extended to the case that the events U_i of a sample set are not a priori equiprobable but have initially given probabilities which we now shall call *a priori probabilities* P_i^0. These may be obtained by prior observations. In this situation a new knowledge of some mean values V^{σ} of random quantities may by gained by new observations. V_i^{σ} may be the values of these variables in the pure states U_i. The question now is, which is the adequate probability distribution P

after accepting this new knowledge? In this case the unbiased guess lies in seeking that P which belongs to minimum information gain

$$K(P, P^0) = \sum_i P_i \ln(P_i/P_i^0) = \min \tag{3.1}$$

under the conditions

$$\sum_i V_i^\sigma P_i = V^\sigma, \tag{3.2}$$

$$\sum_i P_i = 1. \tag{3.3}$$

With *Lagrange* multipliers ζ_σ, Ξ, for infinitesimal δP_i this yields

$$\sum_i (\ln P_i - \ln P_i^0 - \Xi + \zeta_\sigma V_i^\sigma) \delta P_i = 0, \tag{3.4}$$

and finally

$$P_i = P_i^0 \exp(\Xi - \zeta_\sigma V_i^\sigma). \tag{3.5}$$

Generally

$$K(P, P^0) = I(P) - I(P^0) - \sum_i (P_i - P_i^0) \ln P_i^0. \tag{3.6}$$

If P^0 is a generalized canonical distribution

$$P_i^0 = \exp(\Psi^0 - \lambda_\nu^0 M_i^\nu), \tag{3.7}$$

we obtain

$$K(P, P^0) = I(P) - I(P^0) - \lambda_\nu^0 (M^\nu - M^{0\nu}). \tag{3.8}$$

In the last term only the mean values of such quantities M_i^ν occur which yielded the distribution P^0 and which are not changed by the new knowledge. This term thus vanishes. In this case

$$K(P, P^0) = I(P) - I(P^0), \tag{3.9}$$

the information gain is equal to the increase of the *Shannon* information caused by the observation of the new values V^σ. This illustrates once more the concept of information gain.

So far we considered the case that the new knowledge was given by mean values V^σ not belonging to the set which had determined P^0. We can, however, include the case that some of the M^ν are "corrected" by the new observation. In this case we define the deviations of the corrected M^σ from their values in P^0 to be some V^σ themselves and introduce the uncorrected mean values $M^{0\nu}$ into P as parameters.

We can express the information gain

$$K(V) = \Xi(\zeta) - \zeta_\sigma V^\sigma \tag{3.10}$$

as a function of the set V and show that it is the *Legendre* transform of $\Xi(\zeta)$. For this purpose we have to prove that

$$\frac{\partial \Xi}{\partial \zeta_\sigma} = V^\sigma. \tag{3.11}$$

This can be done easily by considering another information gain:

$$K(P + \delta P, P) = \delta \Xi - (V^\sigma + \delta V^\sigma)\delta \zeta_\sigma. \tag{3.12}$$

This equation is a consequence of

$$K(P + \delta P, P) = \delta K(P, P^0) - \sum_i \delta P_i \ln(P_i/P_i^0), \tag{3.13}$$

where

$$\delta K(P, P^0) = \delta \Xi - \zeta_\sigma \delta V^\sigma - (V^\sigma + \delta V^\sigma)\delta \zeta_\sigma. \tag{3.14}$$

As the expression in eq. (3.12) never is negative, we indeed obtain eq. (3.11). Now the *Legendre* transform expressed by eq. (3.10) yields

$$\frac{\partial K}{\partial V^\sigma} = -\zeta_\sigma. \tag{3.15}$$

It is obvious that the results of this chapter are true also if the events of the sample set are continuous. Then eq. (3.5) has to be replaced with

$$\rho(x) = \rho^0(x) \exp(\Xi - \zeta_\sigma V^\sigma(x)). \tag{3.16}$$

We can interprete the general relation of eq. (3.9) also in a form as follows. By expressing I as function of the set M and using eq. (2.16) we rewrite eq. (3.9) as

$$K(P, P^0) = I(M) - I(M^0) - \left(\frac{\partial I}{\partial M^\nu}\right)^0 (M^\nu - M^{0\nu}). \tag{3.17}$$

This means that K is this part of an expansion of I with respect to powers of $M^\nu - M^{0\nu}$ which remains if we drop the linear term. We can say that K is the "nonlinear" part of this power expansion.

2   Thermodynamics of Equilibria

In this section of the book we shall be concerned with the theory of thermo-dynamic equilibrium states. This theory is often called "thermostatics" because it does not describe the change of a physical system with time. The question how fast processes occur is not included. On the other hand, the direction of time, the distinction of the future from the past, is important in this part of thermodynamics. The existence of irreversible processes which only can occur in the direction of the future is typical for thermodynamics and makes the question important which changes from one equilibrium to another one are possible and which are not. In the light of this question it does not seem unjustified to retain the traditional name of equilibrium "thermodynamics".

Historically, thermodynamics first arose as a part of physics using special in-herent means to describe states of physical systems. Characteristic states were defined by particular macroscopic measuring processes which are applicable only to this particular class of states, the so-called *thermal states*. To give an example, a typical process of this kind is the measuring of temperature. It requires an equilibrium of the thermometer with the observed system in a sufficiently long time to reach an "equilibrium state" which then remains unchanged. Such a very specific state is a *thermal state*, only in which the temperature is defined. At this level we call the description *phenomenological* or *macroscopic* thermodynamics. The last name is used in contrast to a *microscopic* or *statistical* thermodynamics which seeks a foundation of the phenomenological theory on the basis of physical laws about the motion of molecules or other *microscopic* components of the system. These basic laws are the laws of mechanics or laws of dynamics of radiation fields in classical or in quantum theory.

First we shall develop the concept of thermal states and its connection with a statistical description of *microstates*. Then we shall deduce laws of the phenomenology from this connection. The phenomenological theory has an impressive logical structure and is a useful tool which can be applied also without the knowledge of the statistical theory. It is a closed theoretical system and will be presented in its basic concepts in subsection 2.3. There we shall start from basic laws which had already been found before a statistical foundation was known. This statistical foundation will be part of sub-section 2.2.

In subsection 2.4 we shall discuss phenomena which are typical for the inter-ference of quantum mechanics. They are observed preferably at low temperatures. Indeed the development of quantum mechanics was closely connected with low tem-perature physics.

2.1   Thermal States

Phenomenological thermodynamics is restricted to the description of large systems which are composed of many particles or quanta. As already pointed out, it is restricted to special states, the so-called *thermal states*. These states can be described by relatively few macroscopic *thermal variables* like temperature, pressure, or density of matter. To begin with, *thermal equilibria* are such states. They do not change with time in the macroscopic observation, as long as the environment remains unchanged. Such an equilibrium state will be assumed by the system after a sufficiently long time when the system is brought into an environment which is in a thermal state itself. For instance, we can sufficiently describe an equilibrium state of a gas in a vessel by the two variables temperature and volume, whereas in general a nonequilibrium state requires far more macroscopic variables, like density and flow velocity in different space points. It is characteristic for the problems connected with a systematic description on the phenomenological level that we already used the concept of the thermal state for the environment when we tried to express what a thermal state of the system is.

The description by thermal variables is not restricted to genuin thermal equilibria. It is applicable also to nonequilibrium states which do not change too fast macroscopically. They have to be "slow" in the following sense we shall explain again by the example of measuring temperature as a typical thermal variable. The measuring requires a long enough time for the mercury column or another device to come of rest. This time, however, has to be short on the time scale on which the comparable "slow" changes of the system itself are observed. We call this situation a *time scale separation* of these slow changes on a *long time scale* from the fast equilibration of the thermo-meter on the *short time scale*. Such time scale separations are typical necessary conditions for the applicability of thermodynamics to nonequilibrium processes.

A central theme of this book is the connection between the *macrostates* and the *microstates* of a system. The macrostates are defined by relatively few macroscopic thermal variables; the microstates, however, are defined by values of the enormous number of the dynamical variables of all particles or quanta the system is composed of. Correspondingly, an enormous number of microstates can be possible in one macro-state. We shall be confronted with the central question of the probabilities of the micro-states in a given macrostate, i.e. in a thermal state.

In the first chapter of this subsection we shall discuss the role of the direction of time in thermodynamics, the discrimination of future compared to past. In the second chapter we shall consider classical mechanics of ensembles. The fourth and the more special fifth chapter are devoted to the generalization to quantum mechanics. In the third and the sixth chapter we shall be concerned with the construction of prob-ability distributions over microstates and their connection with thermal variables.

2.1.1   Direction of Time

One of the fundamental questions for the interpretation of thermodynamics is how the macroscopic thermodynamical laws can be irreversible with respect to time whereas we wish to explain them by reversible laws on the *microscopic* level of micro-states. Let us consider a *time reversal*. This is the operation by which the time t is repaced with $-t$. Such an operation is realized by leaving a moving picture run backwards. We call dynamical laws *reversible* if any process following this dynamics is changed into a possible process by a time reversal. In thermodynamics we know processes which do not possess this property. They are *irreversible*. Such a process is plain heat conduction for instance, which goes only from hot to cold and not into the opposite direction, or plain diffusion which goes only from a higher to a lower concentration in a dilute solution.

Very generally we can say, that statistical physics is concerned with probabilities of certain states of a physical system if some knowledge about the dynamics of the system and the values of macroscopic thermal variables are given. In terms of probability theory, this knowledge represents a conditional event, say c. The state whose probability is asked for is the consecutive event, say ξ. We shall be concerned, for instance with the probability of a microstate ξ. As a rule, the event ξ occurs at a later time than c. We designate the probability that ξ takes place at time t if c was given at time zero by

$$P\begin{pmatrix} \xi & c \\ t & 0 \end{pmatrix}. \tag{1.1}$$

We call this a *progressive* probability P_+ if $t > 0$. The probabilities in conventional thermodynamics are of this kind. In special scientific problems, however, the *regressive* probability P_- can be of interest, for which $t < 0$. Then the probability of an event in the past is asked for. The conditional event is an observation made at present time. For instance, in cosmology we are often interested in unobserved earlier states of astrophysical objects. The sequence of conditional and consecutive event is the sequence of the inductive implication yet not always of time.

Statistical mechanics as the foundation of phenomenological thermodynamics uses progressive probabilities when asking what is to be expected on account of the last observation. By this question the direction into the future is distinguished from the direction into the past. Therefore we are faced with the fact that the results of the theory show this distinction in the form of irreversible laws.

A fundamental property of an adequate description by a statistical theory has to be that the information about the question which microstate a system is in cannot increase with time after the last observation if the system has been isolated in the mean time. "Isolated" means that it has not been influenced by another system. This law is irreversible as it distinguishes future against past. If P_+ is the probability distribution over the microstates of the system, this means for the *Shannon* information $I(P_+)$ at two times $t_1 < t_2$:

$$I_1 \geqslant I_2. \tag{1.2}$$

This inequality which is not symmetric with respect to time reversal is wrong if $t_1 > t_2$.

If we assume reversibility of microscopic dynamics, we obtain the equation

$$P_+\begin{pmatrix} \xi & c \\ t & 0 \end{pmatrix} = P_-\begin{pmatrix} \xi & c \\ -t & 0 \end{pmatrix}, \tag{1.3}$$

which is defined only for $t > 0$. This equation, however, does not give rise to a reversibility of macroscopic laws. If we measure the knowledge about unobserved states ξ in the past by a smooth continuation of $I(P_+)$ (i.e. with a steady time derivative) to negative t, this information will never increase with decreasing time, in contrast to $I(P_-)$. Yet the information $I(P_-)$ belongs to the regressive probability, not used in the statistical theory which provides the foundation of conventional thermodynamics. This theory is based on the use of progressive probabilities.

As a rule, the dependence of P_+ on the condition c is not expressed explicitly in the usual notation and thus veiled if the change of $P_+(t)$ with time t is described by dynamical equations, which in general are irreversible. The analytical continuation of the solution $P_+(t)$ to negative values t makes no sense if the observation of c was made at time zero. The continuation has nothing to do with conclusions about earlier states, occurring before the last observation.

In all of the following chapters we shall be concerned with progressive probabilities if the contrary is not emphasized expressively. Therefore we shall designate them simply by P and drop the subscript $+$ in conformity with the usual notation.

To give an illustration of the progressive character of thermodynamic probabilities we consider a system in a thermal equilibrium. In this macroscopic state *fluctuations* of quantities from their mean values are possible and happen at random. If we exclude phase transitions, such fluctuations are "small" but can be observed on a more detailed but nevertheless macroscopic level. Such a level is sometimes called a *mesoscopic* level between macro- and microscopic description. The fluctions are random quantities with the mean value zero. Once, however, a nonzero fluctuation of a quantity is observed, we may ask for the progressive conditional probability of the future values of the fluctuation. To this probability corresponds a conditional mean value of the fluctuation dependent on time. As a rule, the absolute value of this mean value will not increase in time if no other observation has been made in the mean time. If we describe the change of this mean value with time t by a function $V(t)$, we can apply the description only at positive t in the case that the last observation took place at time zero. An analytical continuation of $V(t)$ to negative t has nothing to do with the behavior of the fluctuation. Neverthheless it is quite adequate to ask for the behavior of the fluctuation before time zero. In the case that the microdynamics is reversible, then $V(|t|)$ with negative time will describe the regressive mean value of the fluctuation. As a rule this $V(|t|)$ is not the analytical continuation of $V(t)$ to negative time.

2.1.2   The *Liouville* Theorem

A macroscopic state in thermodynamics can be described by a set of variables which in the interpretation of statistical mechanics are identified with mean values of microscopic quantities occurring in microdynamics. Thus the question arises about the probability distribution of the microstates corresponding to a given macrostate. To find the distribution by the method of unbiased guess, requires that the set of microstates be a set of "a priori" equally probable events. In *classical mechanics* the microstates of the molecules, or other elementary components, are continuous events and have to be mapped by a continuous parameter space. In the following we shall show that a parametrization of this space exists for which the assumption of the a priori equal probability of volumes of equal size is supported by the structure of classical mechanics. The support is given by the so-called *Liouville* theorem which will be derived as follows.

We shall consider a physical system which satisfies the dynamics of classical mechanics. A dynamical state of such a system can be described by the set of say f independent generalized coordinates q_k of configuration and their canonical conjugate momenta p_k. The subscript k runs from 1 to f. We call f the number of "degrees of freedom". The whole set of all q_k and of all p_k will be called.

$$\xi = (q_1, \ldots, q_f, p_1, \ldots, p_f) = (q, p). \tag{2.1}$$

The elements of this set will be designated uniformily by ξ_s, where the subscript s runs from 1 to $2f$. These dynamical states described by ξ are the microstates in statistical mechanics. Their space, the $2f$-dimensional space ξ of the dynamical variables ξ_s is conventionally called the Γ-*space*. Under very general conditions which are assumed to be fulfilled for all isolated systems, there is a *Hamiltonian*

$$H(q, p) \equiv H(\xi) \tag{2.2}$$

and the dynamics is described by the canonical equations

$$\dot{q}_k = \frac{\partial H}{\partial p_k}, \qquad \dot{p}_k = -\frac{\partial H}{\partial q_k}. \tag{2.3}$$

We shall generally designate the time derivative by a dot above the symbol of a quantity. The canonical equations determine the motion $\xi(t)$ of the mapping point in Γ-space for a given initial state $\xi(0)$. The character of this motion becomes easier to survey if *Euklidean* metrics of the Γ-space is introduced, that shall always be done in future. The velocity of this mapping point constitutes a $2f$-dimensional vector field in Γ-space

$$V = (\dot{q}, \dot{p}) = \left(\frac{\partial H}{\partial p}, \ -\frac{\partial H}{\partial q} \right). \tag{2.4}$$

This has as divergence zero,

$$\mathrm{div}\, V = \sum_k \left(\frac{\partial \dot{q}_k}{\partial q_k} + \frac{\partial \dot{p}_k}{\partial p_k} \right) = 0 \tag{2.5}$$

as a consequence of eqs. (2.3).

Let us now consider a probability distribution over the microstates. This is described by a probability density $\rho(\xi)$ in Γ-space which in general will change with time. To ellucidate this motion, we interprete $\rho(\xi)$ by relative frequencies. Then $\rho(\xi)$ is the density of mapping points of an ensemble of many equal systems. These points move in Γ-space with $\xi(t)$. Thus they flow like a fluid with the velocity field $V(\xi)$. During this motion no mapping point vanishes or springs up. Therefore the "continuity equation"

$$\frac{\partial \rho}{\partial t} + \mathrm{div}\,(\rho V) = 0 \tag{2.6}$$

holds. In this equation the derivative with respect to time t is formed on condition that the value ξ be fixed. We may ask about the change of the density ρ as seen by an observer moving with the flow, i.e.

$$\frac{d\rho}{dt} = \frac{\partial \rho}{\partial t} + \sum_s \frac{\partial \rho}{\partial \xi_s}\, V_s. \tag{2.7}$$

The last term on the right hand side of this equation is equal to the divergence of the flow density ρV because of eq. (2.5), and we obtain

$$\frac{d\rho}{dt} = 0. \tag{2.8}$$

This means that for the moving observer, ρ is unchanged by the flow. This result is called the *Liouville theorem*.

The mapping points of the ensemble with the density ρ flow in Γ-space like particles in an incompressible fluid.

> The same mapping points always fill up a volume in Γ-space of equal size.

Not only this fluid as a whole but each arbitrarily small part does not change its volume. As the density of the fluid is allowed to be inhomogeneous in space, it may be more distinct to say that the fluid is "locally incompressible".

It should be stressed that this theorem does not preclude that the shape of the area filled up by the same ensemble points can change dramatically. To show this, a possible change of a small cell is demonstrated in Figure 6. Thus a compact area can diffuse, for instance into a "foamy" one filled with empty holes. For an observer looking only casually and not discerning such "bubbles" in the foam, the density of the ensemble can be very different from that of the compact fluid forming the walls of the bubbles. Only the latter is unchanged on account of the theorem, not the density the observer sees.

If the density does not depend on time, that means if

$$\frac{\partial \rho}{\partial t} = 0, \tag{2.9}$$

then the flow velocity V and the gradient of ρ in the Γ-space have to be orthogonal wherever the gradient is nonzero.

Figure 6 A change of a phase space volume element allowed by the *Liouville* theorem

2.1.3 Equilibrium Distributions

We shall be concerned with the question which probability distribution $\rho(\xi)$ in Γ-space has to be associated with a thermal equilibrium, a "thermal state". This distribution does not depend on time. In the beginning of this section, characteristic properties of thermal states were explained in the framework of the macroscopic "phenomenological" description. Thermal states are described by so-called "thermal variables". A special type of these variables are statistical averages M^ν of quantities which have well defined values $M^\nu(\xi)$ in any microstate ξ of the Γ-space. Such variables are for instance the energy of a system in a heat bath, the volume of a gas in a movable piston, or the number of molecules of a special chemical species in a mixture. Generally we call functions of ξ "phase space functions" or simply "phase functions". We suppose that the quantities M^ν are independent of one another in the sense that none of them is uniquely determined if the others are given. We suppose that a set of macroscopic thermal variables of this kind determines the thermal state uniquely.

The statistical averages we observe are time averages

$$\overline{M^\nu} = \frac{1}{T} \int\limits_0^T dt\, M^\nu(\xi(t)) \tag{3.1}$$

belonging to an observation time T, which is very long compared to the times of microscopic changes. Therefore the averages are practically constant in the equilibrium and can be assumed to be equal to the mean values

$$M^\nu = \int d\xi\, \rho(\xi)\, M^\nu(\xi), \tag{3.2}$$

the so called "ensemble averages" performed with an adequate probability distribution $\rho(\xi)$, a probability density in the Γ-space, independent of time. $d\xi$ is the $2f$-dimensional volume element in the Γ-space. The question of which probability distribution corresponds to the thermal equilibrium state cannot be answered without certain hypothetical suppositions. We shall present two different ways of finding such a distribution, starting with two different hypothetical assumptions.

The unbiased guess method. This method starts with a fundamental hypothesis of classical statistical mechanics, the

Assumption of equal a priori probability of equal phase volumes.

The assumption is that, if absolutely nothing is known about the state of a physical system, the probability of finding the mapping point ξ of its microstate in a particular part of the Γ-space is proportional to the volume of this part. By that we mean the volume in Γ-space, called the "phase volume" of this part. The *Liouville* theorem represents a very important distinction of the Γ-space, i.e. of the canonical variables in comparison with any other parametrization of the microstates, and it is essential to the assumption about the a priori probability. This assumption is not a stringent consequence of the *Liouville* theorem. If, however, the *Liouville* theorem were wrong, the assumption would also be wrong because the homogeneous a priori density would change into an inhomogeneous density because of the dynamics of a conservative system. So we can say that the *Liouville* theorem is an important support for the assumption.

If the mean values M^ν are given, the unbiased guess leads us to the distribution

$$\rho(\xi) = \exp(\Psi - \lambda_\nu M^\nu(\xi)). \tag{3.3}$$

For a given Γ-space, the guess is only based on the knowledge of the mean values M^ν and of the Γ-space. It does not take into account any knowledge about the dynamics in this space. During a certain time at least, there will be practically no influence of the environment on the thermal equilibrium state of a macroscopic system. During this time, the system behaves like an isolated system, and the *Liouville* theorem holds. $\rho(\xi)$ of eq. (3.3) has to be independent of time, satisfying eq. (2.9). According to the context of eq. (2.9), this is fulfilled if the phase functions $M^\nu(\xi)$ are constants of the motion, at least approximately during this time. Correspondingly they are called "conservative" or "quasi conservative".

The ergodic hypothesis. We shall now present an alternative way to derive the equilibrium distribution $\rho(\xi)$. This deduction is based on the so-called *ergodic hypothesis*, which is considerably older than the information theory.

In the conservative system, the *Hamiltonian $H(\xi)$* is a constant of the motion. Therefore the mapping points in the $2f$-dimensional Γ-space always remain at a $(2f-1)$-dimensional surface, the so-called *energy surface*, given by

$$H(\xi) = E \tag{3.4}$$

with a constant value E. Strictly speaking, the Γ-space volume of any super-surface in this space is zero and no finite density $\rho(\xi)$ can be defined for a system which is exactly at the energy surface. Therefore it is more realistic to consider a thin layer at the super-surface with nonzero volume, an *energy shell*, corresponding to the fact that energy E can be determined macroscopically only with a restricted accuracy within certain limits which define the thickness of the energy shell. This thickness shall correspond to an energy difference ΔE smaller than any value that can be registered macroscopically. If the energy of the system is given so sharply that the mapping point is certainly at the energy shell, the probability density in the Γ-space is nonzero only at the shell.

The ergodic hypothesis in its oldest form was the assumption that the mapping point of the system passes through every point of the energy surface during its motion. This version of the hypothesis was soon replaced by the *quasi ergodic hypothesis* which states that the point passes every small cell of the energy shell in a finite time. Today this is simply called *ergodic hypothesis*. After a sufficiently long time, the mapping point of the system will have passed each volume element of the shell. According to the *Liouville* theorem, the time the system was in celles of equal size is the same. For an observer who knows only that the system has an energy value of the shell, but does not know the initial state, the probability density of finding the system in a microstate will be constant at the whole shell. Such a distribution, constant in the energy shell and zero externally, is called a *microcanonical distribution*.

This can be extended to the situation that the system is not isolated but in thermal contact with its environment. Experience shows that a macroscopic system will not change a thermal equilibrium state after being separated by an isolation from the environment. This leads us to the assumption that the probability distribution after the isolation is the same as before. This is allowed if a boundary layer which can be correlated only with the environment is negligibly thin compared to the large system. The system therefore has to be macroscopic by all means.

Now the energy is a random quantity. The conditional probability density in Γ-space, on the only condition that energy has the value of a certain energy shell, is the microcanonical distribution and thus depends only on energy. We have to multiply this conditional density by the probability that the energy has the value of the shell and we obtain the absolute probability density in Γ-space. As both factors are only dependent on the energy, the absolute probability in the Γ-space is a function only of the *Hamiltonian*

$$\rho\,(\xi) = F\,(H\,(\xi)). \tag{3.5}$$

We neglect the interaction energy between the system and its environment. This is in accordance with neglecting the correlations. Then the energy of the larger system that is composed of our system with energy H and its environment with energy H' is the sum of both *Hamiltonians*. Let us call it

$$H''(\xi) = H\,(\xi) + H'(\xi). \tag{3.6}$$

That the system is not correlated with the environment means that

$$F''(H'') = F\,(H)\,F'(H') \tag{3.7}$$

and for the logarithm f of the densities ρ

$$f''(H + H') = f\,(H) + f'(H'). \tag{3.8}$$

This is the same functional equation as eq. (1.4) of chapter 1.2.1 with the solution eq. (1.6). We write the solution in the form

$$f\,(H) = -\,\beta H + \Psi. \tag{3.9}$$

It yields the so-called *canonical distribution*

$$\rho\,(\xi) = (\Psi - \beta H\,(\xi)). \tag{3.10}$$

The parameter β determines the mean value of the energy, and according to the normalization of ρ, the quantity Ψ is a unique function of β. We shall see later that β corresponds uniquely to the temperature. The canonical distribution corresponds to a given mean value of energy, whereas the microcanonical distribution belongs to a sharp energy value.

We consider now the case that the equilibrium is given with the value of further variables M^ν, not only with energy. We may call the energy M^0. All $M^\nu(\xi)$ are constants of the motion. For sharp values of these phase functions, the system can only pass mapping points in the Γ-space which are simultaneously at the different shells belonging to these sharp values. These shells cross one another. Then we have to replace the microcanonical distribution by a density which is constant in the crossing region of the Γ-space and zero on the outside. If, however, the values of the constants of motion $M^\nu(\xi)$ are not given sharply but are random, the equilibrium density $\rho\,(\xi)$ has only to be a function of a linear combination

$$\eta\,(\xi) = \lambda_\nu\,M^\nu(\xi), \tag{3.11}$$

which enters into eq. (3.7) instead of the *Hamiltonian*. So we obtain the generalized canonical distribution eq. (3.3) anew. The special values of the averages M^ν determining the macroscopic thermal states correspond to adequate values of the parameters λ_ν. The quantity Ψ is fixed by the normalization of ρ and is a unique function of the parameters λ_ν. The canonical distribution is a special case of eq. (3.3), in which all coefficients λ_ν are zero with the exception of that of the energy.

***The ergodic theory.** The ergodic hypothesis was introduced by *L. Boltzmann* in 1871. A physical system fulfilling this hypothesis is called *ergodic*, the property *ergodicity*. Already when this concept was first used, it was found that there were several mechanical systems that were not ergodic. They were, however, idealized models and the nonergodicity would be destroyed by the slightest perturbation. It was believed that this was generally the case. Two decades ago it was proved rigorously with examples that this belief was wrong. In 1932 *J. von Neumann* introducted a condition stronger than ergodicity, called the *mixing* character of the system. It is a property of an ensemble of equal systems and it proved necessary to ensure that the ensemble always spreads over the whole energy shell. Any mixing ensemble is ergodic, but an ergodic ensemble is not always mixing. Only few physical systems have been proved to be mixing[5]; the gas of hard spheres is representative. The "ergodic theory" that is concerned with these questions in the framework of general dynamics, not only with regard to the ergodic hypothesis, has developed into an extended field of physical mathematics.

*2.1.4 Statistical Operators in Quantum Mechanics

The state space in quantum mechanics. We shall not restrict our discussion only to classical mechanics as the dynamics of the microstates and thus to statistics in classical Γ-space, but we shall include quantum mechanics as well. The states of a quantum mechanical system in the most detailed description are vectors $|\psi\rangle$ in a *Hilbert* space, the *state space*. An orthogonal basis of this space is composed of all possible states $|\alpha\rangle$ which can be the results of the same *complete measurement*, say A. A complete measurement comprises all simultaneously measurable independent observables. Of course, in a system of macroscopic size it is practically impossible to perform a complete measurement. We are, however, concerned with the question, how physical laws used in the theory of microdynamics are connected with macroscopic laws. The microscopic laws are based on the concept of quantum states. When, in a metaphorical sense, we shall say "measurable" and "observable", we mean that the idealized measurement does not contradict basic principles of quantum mechanics, according to which a simultaneous measurement of certain pairs of quantities is impossible.

On account of the existence of observables which are not simultaneously measurable also in the idealized meaning, the *Hilbert* space has more than one orthogonal basis. The orthogonal states of one basis form an enumerable set. (This is true at least for systems not extending infinitely in ordinary space. Also in infinite space the states can be made enumerable by the introduction of periodic boundary conditions as a mere

---

[5] *V. A. Arnold, A. Avez,* Ergodic Problems of Statistical Mechanics (Benjamin, New York 1968); *Ya. G. Sinai,* Sov. Math. Dokl. **4,** 1818 (1968); Russian Math. Rev. **25,** 137 (1970).

trick of mathematical technic.) Any vector $|\psi\rangle$ and thus any state of the *Hilbert* space can be represented by a linear combination of all the elements $|\alpha\rangle$ of one basis. This means that it can be written in the form

$$|\psi\rangle = \sum_\alpha |\alpha\rangle\langle\alpha|\psi\rangle \tag{4.1}$$

with coefficients $\langle\alpha|\psi\rangle$. In particular this yields for two elements $|\alpha\rangle$, $|\alpha'\rangle$ of the same basis:

$$\langle\alpha|\alpha'\rangle = \delta_{\alpha\alpha'} \tag{4.2}$$

with the *Kronecker* symbol $\delta_{\alpha\alpha'}$. These equations for all elements of the basis are called "orthogonality relations" if α and α' are different, and "normalization relations" if they are equal. It is usual to introduce vectors $\langle\psi|$ of a congruent second *Hilbert* space as *adjoint* to the corresponding $|\psi\rangle$ in a one-to-one mapping. Then $\langle\alpha|\psi\rangle$ is defined as a scalar product of two vectors. In this conventional formalism we can write

$$\sum_\alpha |\alpha\rangle\langle\alpha| = 1 \tag{4.3}$$

with summation over all elements $|\alpha\rangle$ of one complete basis. Then eq. (4.1) can be read as the application of the identity operation 1 to $|\psi\rangle$. For the adjoint vector we write

$$\langle\psi| = \sum_\alpha \langle\psi|\alpha\rangle\langle\alpha| \tag{4.4}$$

and apply thus the operator 1 from the right hand side. The adjunction of the vector $\langle\psi|$ to $|\psi\rangle$ is made an *Hermitian adjunction* by requiring that

$$\langle\psi|\alpha\rangle = \langle\alpha|\psi\rangle^* \tag{4.5}$$

where the asterisk designates "complex conjugate".

The value

$$|\langle\alpha|\psi\rangle|^2 = \langle\psi|\alpha\rangle\langle\alpha|\psi\rangle = P_\alpha \tag{4.6}$$

is interpreted to be the probability of finding the result $|\alpha\rangle$ of the complete measurement A if this is performed in the state $|\psi\rangle$. The state $|\psi\rangle$ is itself the result of another preceding complete measurement Ψ which as a rule is different from A. Thus $|\psi\rangle$ comprises the knowledge we had obtained by this preceding measurement. The knowledge is to be replaced with the new result $|\alpha\rangle$ of the following measurement A. We express this by saying that the state $|\psi\rangle$ is changed into $|\alpha\rangle$ by the measurement A. The change into a particular $|\alpha\rangle$ occurs with the probability P_α of eq. (4.6).

Observables. As already stressed, in general a measurable quantity X called *observable* cannot be measured simultaneously with another arbitrary observable even in the idealized sense of "measuring" we spoke of. The measurement of X however, can always be understood as part of a complete measurement, say A with the corresponding possible results $|\alpha\rangle$. To each $|\alpha\rangle$ belongs a value X_α as the corresponding measured value of X. We call X_α an *eigenvalue* of X to the *eigenstate* $|\alpha\rangle$. If the system has been in the state $|\psi\rangle$ when the measurement is performed, the result $|\alpha\rangle$ will occur with the probability P_α of eq. (4.6). Therefore the mean value of all possible results is

$$\langle X \rangle = \sum_\alpha P_\alpha X_\alpha = \sum_\alpha \langle \psi | \alpha \rangle X_\alpha \langle \alpha | \psi \rangle, \tag{4.7}$$

which is likewise called *expectation value* of X. The observable X can be represented by the operator

$$\mathbf{X} = \sum_\alpha |\alpha\rangle X_\alpha \langle \alpha|. \tag{4.8}$$

The expectation value then is

$$\langle X \rangle = \langle \psi | \mathbf{X} | \psi \rangle. \tag{4.9}$$

Generally an *operator* Ω in *Hilbert* space, which we shall desginate with a bold letter, transforms any vector $|\psi\rangle$ into another one $|\varphi\rangle$:

$$\Omega | \psi \rangle = | \varphi \rangle. \tag{4.10}$$

The observables in particular are *Hermitian* operators, this means that they are linear, have real eigenvalues, and orthogonal eigenvectors. We can express $|\psi\rangle$ by eq. (4.1), can do the same with the left hand side of eq. (4.10), and obtain for a linear operator Ω

$$\Omega | \psi \rangle = \sum_{\alpha\alpha'} |\alpha'\rangle \langle \alpha' | \Omega | \alpha \rangle \langle \alpha | \psi \rangle. \tag{4.11}$$

As this is valid for any $|\psi\rangle$, we can always write Ω in the form

$$\Omega = \sum_{\alpha\alpha'} |\alpha'\rangle \langle \alpha' | \Omega | \alpha \rangle \langle \alpha |, \tag{4.12}$$

where $\langle \alpha' | \Omega | \alpha \rangle$ is called a *matrix element* of the operator Ω in the basis $|\alpha\rangle$. It should be mentioned that the eq. (4.12) is obtained by applying the unit operator eq. (4.3) on both sides of Ω. In any basis an operator is represented by a *matrix*. An observable, as any *Hermitian* operator $\mathbf{X}$ is represented in its own eigenbasis, i.e. in the basis of its eigenvectors, by a diagonal matrix. The eigenvalues X_α are the diagonal elements;

$$\langle \alpha' | \mathbf{X} | \alpha \rangle = X_\alpha \delta_{\alpha\alpha'}. \tag{4.13}$$

Mixture states. As introductory example we first consider the following special situation. The complete measurement A was performed in the state $|\psi\rangle$, but its result is not registered and not known. We know, however, that the system is with the probability P_α in one of the states $|\alpha\rangle$. The expectation value of an arbitrary observable $\mathbf{Y}$ is

$$\langle Y \rangle = \sum_\alpha P_\alpha \langle \alpha | \mathbf{Y} | \alpha \rangle. \tag{4.14}$$

It is usual to introduce the *Hermitian* operator

$$\rho = \sum_\alpha |\alpha\rangle P_\alpha \langle \alpha|. \tag{4.15}$$

The successive application of two operators, say first $\mathbf{Y}$ and then ρ, is called the *product* $\rho\mathbf{Y}$. In our case we can write

$$\rho\mathbf{Y} = \sum_{\alpha\alpha'} |\alpha\rangle P_\alpha \langle\alpha| Y|\alpha'\rangle \langle\alpha'|. \tag{4.16}$$

Generally the sum of the diagonal elements of an arbitrary operator, say Ω, is independent of the special basis and called the *trace* of Ω

$$\operatorname{tr} \Omega = \sum_\alpha \langle\alpha|\Omega|\alpha\rangle. \tag{4.17}$$

We see that

$$\langle\mathbf{Y}\rangle = \operatorname{tr}(\rho\mathbf{Y}). \tag{4.18}$$

The normalization condition of the distribution ρ can be considered to be the special case of eq. (4.18):

$$\operatorname{tr}\rho = 1. \tag{4.19}$$

The trace of a product of two operators is independent of the sequence of the factors. Thus we can commute ρ and $\mathbf{Y}$ in eq. (4.18).

Now we come to the introduction of the general statistical operator. The elements $|\alpha\rangle$ of a basis as results of a complete measurement form a sample set of events. In certain situations, these elements can become random events. Whenever a probability distribution P_α over this set is given, we can define a so-called *statistical operator* ρ by eq. (4.15). Heretofore we demonstrated only one particular situation leading to probabilities P_α of this kind. These probabilities, however, can be obtained in different ways, for instance by an adequate theory. A statistical operator *(density matrix)* describes a state of knowledge of the observer. Nevertheless, it is usual to say that ρ corresponds to a *mixture state* of "the system" in contrast to the *pure states* $|\psi\rangle$. Both kinds of "states" can be described uniformly by operators if we introduce the *projection operator* or *projector*

$$\Lambda_\psi = |\psi\rangle\langle\psi| \tag{4.20}$$

for each pure state $|\psi\rangle$. By use of these operators we can write

$$\rho = \sum_\alpha P_\alpha \Lambda_\alpha \tag{4.21}$$

$$\mathbf{X} = \sum_\alpha X_\alpha \Lambda_\alpha \tag{4.22}$$

$$1 = \sum_\alpha \Lambda_\alpha. \tag{4.23}$$

The probability of finding the result $|\alpha\rangle$ in state $|\psi\rangle$ can be written

$$|\langle\alpha|\psi\rangle|^2 = \operatorname{tr}(\Lambda_\alpha \Lambda_\psi). \tag{4.24}$$

As we can see, it makes no difference if we exchange $|\alpha\rangle$ with $|\psi\rangle$. Therefore

$$\mathrm{tr}\,(\rho\,\Lambda_\psi) = \sum_\alpha P_\alpha\,\mathrm{tr}\,(\Lambda_\alpha\,\Lambda_\psi) \tag{4.25}$$

is the probability of finding the pure state $|\psi\rangle$ in the mixture state ρ by a corresponding complete measurement with the possible result $|\psi\rangle$.

Information measures. As the eigenvalues P_α of ρ form a probability distribution over the sample set of the pure states Λ_α, the *Shannon* information of the statistical operator ρ can be defined by $I(P)$. This means:

$$I(\rho) = \sum_\alpha P_\alpha \ln P_\alpha = \mathrm{tr}\,(\rho \ln \rho). \tag{4.26}$$

In this expression we use the general definition of a function $f(\mathbf{X})$ of any *Hermitian* operator $\mathbf{X}$ of eq. (4.22) to be the operator

$$f(\mathbf{X}) = \sum_\alpha f(X_\alpha)\,\Lambda_\alpha. \tag{4.27}$$

The operator

$$\mathbf{b} = -\ln\rho \tag{4.28}$$

may be called "bit number" in quantum mechanics.

The generalization of *information gain* to quantum mechanics is

$$K\,(\rho,\rho') = \mathrm{tr}\,[\rho\,(\ln\rho - \ln\rho')]. \tag{4.29}$$

The argument for this generalization is a bit lengthier. In general the eigenstates Λ'_β of

$$\rho' = \sum_\beta P'_\beta\,\Lambda'_\beta \tag{4.30}$$

form another orthogonal basis different from that of the eigenstates of ρ. With the probabilities

$$Q_\beta = \mathrm{tr}\,(\rho\,\Lambda'_\beta) = \sum_\alpha \Theta_{\beta\alpha}\,P_\alpha, \tag{4.31}$$

where the matrix

$$\Theta_{\beta\alpha} = \mathrm{tr}\,(\Lambda'_\beta\,\Lambda_\alpha) \tag{4.32}$$

describes transition probabilities, the equation

$$K\,(\rho,\rho') = I(P) - \sum_\beta Q_\beta \ln P'_\beta \tag{4.33}$$

holds. $-\ln P'_\beta$ is the bit-number of the probability P'_β of Λ'_β in state ρ'. In state ρ, however, Λ'_β has the probability Q_β. Therefore the second term of the right hand side in the last equation is the mean value of the mentioned bit-number in the state ρ. Thus in quantum mechanics, eq. (4.33) is the adequate generalization of the *Kullback* measure of two probability distributions P, P', i.e.

$$K(P, P') = I(P) - \sum_i P_i \ln P_i',$$

(4.34)

where the last term is the mean value of the bit number of the probabilties P_i' formed with the distribution P.

Like the ordinary *Kullback* measure, the quantum mechanical $K(\rho, \rho')$ is also never negative. In order to show this, we write it in the form

$$K(\rho, \rho') = I(P) - I(Q) + K(Q, P').$$

(4.35)

The equations

$$\sum_\beta \Theta_{\beta\alpha} = \sum_\alpha \Theta_{\beta\alpha} = 1$$

(4.36)

define $\Theta_{\beta\alpha}$ as a bistochastic matrix. This matrix transforms the distribution P into Q. Therefore, with eq. (3.14) of chapter 1.2.3

$$I(P) - I(Q) \geqslant 0.$$

(4.37)

As $K(Q, P)$ is never negative, the same is true for $K(\rho, \rho')$.

Analogously to the ordinary *Kullback* measure, the quantum mechanical $K(\rho, \rho')$ also vanishes only if ρ and ρ' are identical. This can be seen in the following way. If $K(\rho, \rho')$ vanishes, the term $K(Q, P')$ in eq. (4.35), as well as the left hand side of eq. (4.37) must vanish separately. The first condition yields that the distributions Q and P' are identical. The second condition then requires that the transformation from P to P', which now means from P to Q, because of eq. (4.31), has to be a pure permutation, i.e.

$$\Theta_{\beta\alpha} = 1 \text{ or } 0.$$

(4.38)

We can choose the arbitrary sequence of the subscripts β for making the permutation to the identity. That means, ρ and ρ' are indeed identical.

Generalized canonical ρ. The method of unbiased guess can also be applied in quantum mechanics to construct the adequate statistical operator ρ to given mean values

$$M^\nu = \text{tr}(\rho M^\nu)$$

(4.39)

of a set of independent observables $\mathbf{M}^\nu$. This is possible if a fundamental principle of quantum statistics is accepted, namely that pure states Λ_α have equal a priori probability. Such a state has to be absolutely "pure" in the sense that the complete measurement, the result of which defines the state, comprises indeed all independent simultaneously measurable quantities. Such a measurement includes for instance not only all position coordinates or all momenta of a many-particle system but all spin variables as well. This principle plays the same role in quantum mechanics as the principle of homogeneous a priori probability density in Γ-space plays in classical mechanics. We can understand the latter as consequence of the quantum mechanical principle by considering quantum mechanics as the fundamentals of classical mechanics. Because of the uncertainty relation

$$\Delta p_k \, \Delta q_l = \hbar \delta_{kl} \tag{4.40}$$

all pure states require the same volume in Γ-space, proportional to $\hbar^f$. Here and in the following we desidnate *Planck's* constant h divided by 2π with $\hbar$. Based on the

> Principle of equal a priori probability of pure states,

the adequate ρ to given mean values eq. (3.39) is the one with minimum information $I(\rho)$. Thus we have to require

$$I(\rho) = \operatorname{tr}(\rho \ln \rho) = \min, \tag{4.41}$$

$$\operatorname{tr}(\rho \mathbf{M}^\nu) = M^\nu, \tag{4.42}$$

$$\operatorname{tr}\rho = 1. \tag{4.43}$$

By use of *Lagrange* multipliers λ_ν, $-(\Psi + 1)$ this yields:

$$\operatorname{tr}\left[(\ln\rho - \Psi + \lambda_\nu \mathbf{M}^\nu)\delta\rho\right] = 0. \tag{4.44}$$

We can now conclude in the same way as was explained in chapter 1.3.1 following eq. (1.15). To begin with, the multipliers are free, yet $\delta\rho$ is not. The number of restrictive equations (4.42, 4.43) is equal to the number of *Lagrange* multipliers. Putting the bracket equal to zero leads us to

$$\rho = \exp(\Psi - \lambda_\nu \mathbf{M}^\nu), \tag{4.45}$$

where the conditions eq. (4.42, 4.43) have to be satisfied by the adequate choice of the *Lagrange* parameters the number of which is equal to that of the conditions. Not only the probability distribution P_α in eq. (4.15) but also the basis $|\alpha\rangle$ then is determined. It should be mentioned that, according to the given deduction, it is permissible that the observables $\mathbf{M}^\nu$ not be simultaneously measurable. This means that they may be non-commutable with one another. As the exponent with real *Lagrange* multipliers is an *Hermitian* operator the exponential function in eq. (4.45) is defined according eq. (4.27). So far this holds for the result of the unbiased guess based on instantaneous expectation values M^ν. If, however, ρ has to describe a thermal equilibrium, the analogue is true that was said for the classical equilibrium distribution. Then the observables $\mathbf{M}^\nu$ have to be constants of the motion, at least in good approximation. They have to be conservative or at least quasi conservative quantities. This means practically that they have to commute with the *Hamiltonian*.

We shall call ρ of eq. (4.45) a *generalized canonical statistical operator* or a *generalized canonical distribution*. It leads us to the equations of chapter 1.3.2 between λ_ν and the mean values M^ν in the same way they were obtained for ordinary probability distributions. Thus, without any change, these relations are valid in quantum mechanics too. They are consequences of the relations

$$I(\rho) = \Psi - \lambda_\nu M^\nu, \tag{4.46}$$

$$K(\rho + \delta\rho, \rho) = \delta\Psi - \delta\lambda_\nu(M^\nu + \delta M^\nu) \geqslant 0. \tag{4.47}$$

These consequences will turn out to be determining the basic structures of macroscopic thermodynamics, independently of whether classical mechanics or quantum mechanics is the microscopic basis.

The result of an unbiased guess for an *additional knowledge*, as described in chapter 1.3.3, can also be directly generalized to quantum mechanics with only one exception. Instead of eq. (3.14) of the mentioned chapter, we have to write correctly

$$\rho = \exp\left(\ln \rho^0 + \Xi - \zeta_\sigma V^\sigma\right) \tag{4.48}$$

because if ρ^0 is not commutable with the last term of the bracket, then this dissimilarity is of importance. It is, however, of no importance if both statistical operators correspond to thermal equilibria and thus all occurring operators are conservative and commutable.

*2.1.5   The *Wigner* Funktion

In the preceding chapter we saw that, in a certain sense, the statistical operator in quantum mechanics plays the role the probability distribution in phase space plays in classical mechanics. If, however, we seek for the quantum mechanical correlate which becomes the classical probability density in the Γ-space with the limiting process which reduces quantum mechanics to classical mechanics, we are led to another quantity, the *Wigner function*[6]. Although this concept is not necessary for the structure of macroscopic thermodynamics, we shall represent it in this chapter for the sake of completeness because it is often used in literature.

We consider a "canonical" system with f canonical configuration coordinates $\mathbf{q}_k$ and their complementary conjugate momenta $\mathbf{p}_k$. They cannot simultaneously have defined values. Therefore we cannot define a probability distribution in the Γ-space of these variables (q, p). It is, however, possible to define a function $F(q, p)$ which in a quantum mechanical mixture state ρ gives the probability density $w(q)$ of the coordinates q and the probability density $u(p)$ of the momenta p in the equations

$$\int dp\, F(q, p) = w(q) \tag{5.1}$$

and

$$\int dq\, F(q, p) = u(p). \tag{5.2}$$

These equations are satisfied in classical mechanics if we replace $F(q, p)$ with the classical probability density in Γ-space.

In an infinite configuration space not only the eigenvalues q_k of the quantum mechanical observables $\mathbf{q}_k$ but also the eigenvalues p_k of the momenta $\mathbf{p}_k$ become continuous. The corresponding eigenvectors $|q\rangle$ form a complete basis of the *Hilbert* space. The eigenvectors $|p\rangle$ form another complete basis. For the continuous sets we have to replace eqs. (4.2, 4.3) with

$$\langle q | q' \rangle = \delta(q - q') \tag{5.3}$$

---

[6] *E. P. Wigner*, Phys. Rev. **40**, 749 (1932).

and

$$\int dq \, |q\rangle\langle q| = 1 \tag{5.4}$$

with the f-dimensional delta-function $\delta(q)$. These equations hold also if we replace q, q' with p, p'.

In the mixture state with the statistical operator ρ the probability densities in the parameter space of the eigenvalues q or p respectively are

$$w(q) = \langle q | \rho | q \rangle, \tag{5.5}$$
$$u(p) = \langle p | \rho | p \rangle. \tag{5.6}$$

With the completeness equation (5.4) of the basis $|q\rangle$ and the corresponding completeness of the basis $|p\rangle$, it becomes evident that

$$F(q, p) = \langle p | \rho | q \rangle \langle q | p \rangle \tag{5.7}$$

fulfills the required eqs. (5.1, 5.2) for $F(q, p)$. There also exists, however, another solution

$$F^*(q, p) = \langle q | \rho | p \rangle \langle p | q \rangle, \tag{5.8}$$

which is the complex conjugate of $F(q, p)$. Any linear combination of these two solutions is again a solution. The real one, this is the real part of $F(q, p)$, can become negative, and its use is less advantagenous for the following than that of F or F^*. We shall discuss the use of the *Wigner function* $F(q, p)$.

To any state vector $|\psi\rangle$ in the *Hilbert* space belongs a *Schrödinger* wave function

$$\psi(q) = \langle q | \psi \rangle, \tag{5.9}$$

and it holds that

$$\langle q | \mathbf{p}_k | \psi \rangle = \frac{\hbar}{i} \frac{\partial}{\partial q_k} \, \psi(q). \tag{5.10}$$

If particularly $|\psi\rangle$ is an eigenstate $|p\rangle$ of the whole set $\mathbf{p}$ with the eigenvalues p_k, we obtain

$$\frac{\hbar}{i} \frac{\partial}{\partial q_k} \langle q | p \rangle = p_k \langle q | p \rangle, \tag{5.11}$$

with the result

$$\langle q | p \rangle = C \exp\left(\frac{i}{\hbar} pq\right) \tag{5.12}$$

where

$$pq = \sum_k p_k q_k. \tag{5.13}$$

The constant C is determined by the normalization requirement

$$\langle p' | p \rangle = |C|^2 \int dq \, \exp\left[\frac{i}{\hbar} (p - p') q\right] = \delta(p - p') \tag{5.14}$$

as $h^{-f/2}$ with the number f of degrees of freedom.

To find this result, we used the *Fourier* representation of the one-dimensional *Dirac* delta-function

$$\delta(t) = \frac{1}{2\pi} \int\limits_{-\infty}^{\infty} d\omega \exp(-i\omega t), \tag{5.15}$$

which follows from the general connection between a function $f(t)$ and its *Fourier* transform

$$f(t) = \frac{1}{2\pi} \int\limits_{-\infty}^{\infty} d\omega \, \hat{f}(\omega) \exp(-i\omega t) \tag{5.16}$$

$$\hat{f}(\omega) = \int\limits_{-\infty}^{\infty} dt \, f(t) \exp(i\omega t). \tag{5.17}$$

Due to this, the *Fourier* transform of $\delta(t)$ is 1.

Se we finally obtain

$$\langle q|p\rangle = h^{-f/2} \exp\left(\frac{i}{\hbar} pq\right), \tag{5.18}$$

and

$$\langle q|p\rangle\langle s|p\rangle = h^{-f/2} \langle q+s|p\rangle. \tag{5.19}$$

We are now able to write

$$\int dp \, F(q, p) \langle s|p\rangle = h^{-f/2} \langle q+s|\rho|q\rangle. \tag{5.20}$$

If we introduce q' instead of $s+q$, it yields the so-called "spatial" density matrix

$$\langle q'|\rho|q\rangle = h^{f} \int dp \, F(q, p) \langle q'-q|p\rangle, \tag{5.21}$$

which determines ρ uniquely. We see that this spatial matrix can be obtained by a *Fourier* transformation if the *Wigner* function $F(q, p)$ is given:

$$\langle q'|\rho|q\rangle = h^{f} \int dp \, F(q, p) \exp\left(\frac{i}{\hbar} p(q-q')\right). \tag{5.22}$$

Care has to be taken in the case of a system of equal particles We consider a system of *Bose* particles, the only ones which correspond to classical dynamics. They can assume only states $\langle p|q\rangle$ that are symmetric with respect to any permutation of the particle coordinates q with fixed momenta p. But then our description is not canonical anymore in a rigorous sense, as the complementary relations between q and p and with them eq. (5.18) are changed. A better description would be given by the quantum field theory (in older literatur called "second quantization"). We can, however, use our results if we include into the formation of traces of type

$$\langle A \rangle = \mathrm{tr}\,(\rho\,A) \tag{5.23}$$

a summation over all particle permutations of the q' to fixed q and a division by $N!$ if N is the number of the particles.

2.1.6 Thermal Variables

In the case of quantum mechanics, we have to replace the Γ-space with the *Hilbert* space of pure states as microstates. In the following we shall not mention quantum mechanics separately so long as the results are the same. Only such quantities M^ν are suitable variables which change so slowly with time that their change can be registered by macroscopic means. A particular quantity of this kind is the energy which is conserved in a strictly isolated system. In an open system which can exchange energy with its environment, the larger the system is, the slower the relative change of energy will be. The number of molecules of a special species is conserved in an isolated system as well if no chemical reactions occur during the observation time. Another suitable thermal variable is the volume of the above mentioned piston, the motion of which is slow enough to be observed macroscopically. In an open system such slowly changing quantities are called *quasi-conserved*.

Describing particle numbers as random quantities in such a system requires an extension of the space of microstates. Γ-space and *Hilbert*-space were defined for a system with fixed particle numbers. We can, however, introduce the direct sum of all these spaces with different particle numbers as a higher state space, which in the future we shall call Γ- or *Hilbert*-space respectively, as well, as will be explained in detail in chapter 4.1.2. This extension of the state space is useful because we shall sometimes define a system by all particles in a fixed open volume. The extension is not necessary if the system is defined by a fixed set of particles, maybe in a changing volume.

Statistical physics asks for the probability of the microstates in a thermal state. In a macroscopic system, of course, an enormous number of microstates in possible, since in the thermal state only few macroscopic variables M^ν are given which are interpreted as mean values of microscopic variables. The fundamental principle governing the a priori probability of the microstates allows the unbiased guess which yields the generalized canonical distribution. We shall write the latter in a uniform way common for classical and quantum mechanics:

$$\rho = \exp\,(\Psi - \lambda_\nu M^\nu). \tag{6.1}$$

In classical physics, $\mathbf{M}^\nu$ stands for $M^\nu(\xi)$. Correspondingly, for any function $F(\xi)$ in Γ-space we shall often write $\mathbf{F}$,

$$\langle F \rangle = \int \mathrm{d}\xi\,\rho(\xi)\,F(\xi) = \mathrm{tr}\,(\rho\mathbf{F}). \tag{6.2}$$

A function $F(\xi)$ in Γ-space, which is also called *phase space*, is conventionally called a *phase function*. The normalization

$$\mathrm{tr}\,\rho = 1 \tag{6.3}$$

in particular yields

$$\Psi = - \ln \mathrm{tr} \exp\left(- \lambda_\nu \mathbf{M}^\nu\right). \tag{6.4}$$

Extensive variables. Parts of a physical system situated in different parts of ordinary space can be considered to be physical systems as well. Then most of the above mentioned quasi conserved quantities M^ν are additive with respect to the composition of these parts to the whole system. This is true, for instance, for volume, particle numbers, magnetic moments, and electrical charge. Strictly speaking, it is not true for energy because an interaction energy should be added to the energy of subsystems. The interaction is responsible for exchanges between the subsystems. Nevertheless, the contribution of the interaction energy to the total energy is considered to be negligible on a macroscopic scale in standard thermodynamics. If we consider the limiting process to macroscopic dimensions, this contribution, as a rule, changes in the same order as the interface between the parts, whereas the bulk energy increases in the order of the volume. In this approximation, energy becomes additive as well, as usually assumed in macroscopic thermodynamics. We have to stress that this approximation is justified by the short range of intermolecular forces. The approximation would fail totally in a system of mass points which interact with one another only by gravitation.

In an equilibrium state of a homogeneous system, the additive variables M^ν are proportional to the volume and therefore called *extensive variables* or *extensities*. We include energy.

Intensive variables. For a system composed of two subsystems I and II, in general

$$\mathbf{M}^\nu = (\mathbf{M}^\nu)_\mathrm{I} + (\mathbf{M}^\nu)_\mathrm{II} + (\mathbf{M}^\nu)_\mathrm{I\,II}. \tag{6.5}$$

Here we may include again the occurrence of an interaction term, the last term of the sum. In the generalized canonical distribution of the whole system, all three terms obtain the same coefficient λ_ν. Thus λ_ν does not depend on the extension of the system and is therefore called an *intensive variable* or an *intensity*. We call the intensity λ_ν and the extensity M^ν *thermodynamic conjugates* if they belong to the same index ν. We can characterize the distribution eq. (5.1) and thus the thermal state by the set of intensities λ_ν as well as by the set of the extensities M^ν. (A restrictions has to be made with respect to systems with phase transitions for which additional information is necessary to make the thermal state unique.) We obtain the following important theorem:

> All systems in mutual thermal equilibrium assume equal values of the intensities which are conjugate to independent extensities.

The intensities determine the equilibrium state and therefore we can formulate a further conclusion which is called the *Zeroth Law* of thermodynamics:

> If two systems are in equilibrium with a third one, they are also in equilibrium with each other.

It should, however, be stressed that when we shall call all mean values M^ν "extensities" and all multipliers λ_ν "intensities" in the future, this will be in accordance with a general convention but not with the definition stating that extensities are always proportional to the volume in a homogeneous system. For instance, the surface of a droplet can be a special M^ν as well. Moreover, the use of the term "intensity" is not unique. Also such densities as of matter, energy, or of other quantities M^ν are often called "intensities". Of course, a spatial density is independent of volume, but can be a statistical mean value M^ν in Γ-space. In the following, we shall use the term "extensity" in a more figurative sense for any M^ν occurring in the distribution of eq. (6.1) as mean value of a phase function $\mathbf{M}^\nu$ in classical physics or of a corresponding observable in quantum physics. We then shall call the corresponding multiplier λ_ν the conjugate "intensity".

As already stressed, to obtain the above formulated theorem, we require that the $\mathbf{M}^\nu$ are independent of one another. The surface of a spherical droplet, for instance, depends on the volume. In chapter 2.3.7 we shall see that therefore the intensity "pressure" in the droplet is different from the pressure of the surrounding vapor. Yet this is not a real inconsistency with the general principle of equal intensities in equilibrium because in the droplet the conjugate extensities volume and surface depend on each other. This yields intensities dependent on each other as well.

Inhibited equilibria. As mentioned already at the beginning of this section, the use of thermal variables is not restricted to thermal equilibrium states but is possible also for states changing sufficiently slowly with time. This means that a time scale separation has to be present between the "fast" equilibration to an instantaneous equilibrium and the "slow" change of the macrosopic state.

This separation leads us to the introduction of the concept of a *relative equilibrium* in contrast to the *absolute equilibrium*. The first is a macroscopic state in which parts of the system are in an equilibrium state of their own, yet not in equilibrium with one another. In the absolute equilibrium, all parts are in a common equilibrium. A relative equilibrium is a state of the same kind as if the exchange of energy and matter between the parts were interrupted by isolated walls and therefore each part were in its own equilibrium. Thus such a state is also called an *inhibited* or *frozen* equilibrium. For instance, detonating gas can be kept unaffected in an inhibited equilibrium for a long time without reacting. The reaction is then inhibited. By an ignition or by the addition of a catalyst, the reaction occurs in a vehement detonatio. After a certain time then, the absolute equilibrium may be reached. In this example, the components of the mixture are the "parts" of the system, though they are not separated in space. They were initially separated by the inhibition of the reaction. In a metaphorical sense, this inhibition is the isolating "wall" which is removed by the catalyst. In thermodynamic considerations, the concept of inhibited equilibria is used very frequently.

2.2   Statistical Foundation of the Macroscopic Scheme

The "phenomenological" or "macroscopic" scheme of thermodynamics which originally was developed as an isolated branch of physics based on specific empirical foundations was only at later time connected with first principles of other fields of physics by the statistical interpretation. In a systematic way this was done when phenomenological thermodynamics already had obtained a specific closed structure. Our intention is to present the basic elements of this structure in a concise form in section 2.3. In the present section, however, we shall discuss the statistical foundation of laws and concepts which in the phenomenological scheme were introduced by original postulates.

In the first chapter the *Second Law* of thermodynamics will be considered with respect to the statistical interpretation. This law in particular determines the specific character of thermodynamics. It is closely connected with the concept of *entropy* and with the existence of irreversible processes. We may say, it is the central law in thermodynamics, separated from first principles in other fields of physics.

In the second chapter of this subsection we shall discuss the distinction between heat and work which in phenomenology is a starting point but appears in the statistical theory in a more secondary way. This distinction will be important for the last chapter of this subsection in which the limitations are considered of the possibility to transform heat into work. In comparison to these chapters the two remaining chapters are concerned with problems of more technical character.

2.2.1   The Second Law of Thermodynamics

Entropy. In the conventional terminology the law of energy conservation, if used in thermodynamics, is called the *First Law* of thermodynamics. We shall come to the discussion of this law in the next chapter Here we are concerned with the so-called *Second Law*. It can be formulated in very different ways so that a logical deduction is necessary to show the congruence of the different formulations. Relatively transparent with regard to its consequences is to say that it is the principle of the existence of irreversible processes. In chapter 2.3.1 we shall show that this is totally equivalent with the following formulation:

> For an isolated thermodynamic system a quantity exists uniquely associated with each thermal state which never decreases with time evolution.

This quantity S is called *entropy*. This word is a construction proposed by *R. Clausius*, who introduced this concept into the phenomenological theory around the year 1860. The word was based on Greek $\overset{c}{\eta} \ \tau\rho o\pi\acute{\eta}$ with the meaning "transformation" and was made similar to "energy" by purpose[7]. An "isolated" system is not in interaction with another one. We call a system "thermodynamic" if its state can be described by thermal quantities.

As pointed out in chapter 2.1.1, the thermodynamic statistical description operates with progressive probabilities of events after the last observation. In thermal states, the last observation belongs to relative few mean values M^ν which never can give the precise knowledge of a microstate. Therefore the development of the microstate in time is not exactly known. The description is restricted to probabilites over the microstates. The information associated with their distribution cannot increase after the last observation. In thermal states this information is a function $I(M)$ of the state variables M^ν. Consequently, there exists a function which has the mentioned features of the entropy:

$$S(M) = -kI(M). \tag{1.1}$$

---

[7] *R. Clausius*, Abhandlungen über die mechanische Wärmetheorie (Vieweg, Braunschweig 1867).

This identification of macroscopic entropy with the lack of information about the microstate is the basic link between statistics and phenomenological thermodynamics. The constant k gives the units of entropy which are conventional in phenomenological thermodynamics. It is called the *Boltzmann constant*.

The properties of the generalized canonical distribution, which in this interpretation is comprehended as the consequence of the unbiased guess method, are determining for the fundamental structures of thermodynamics. In particular they give the *Gibbs* fundamental equation (2.16) of chapter 1.3.2, which now can be written in the form

$$dS = k\,\lambda_\nu\,dM^\nu. \tag{1.2}$$

Entropy as an extensity. In chapter 1.2.1 we saw that the information I is additive for uncorrelated subsystem. Due to the identification eq. (1.1), this is true likewise for the entropy S. In the same way as we did when discussing the extensivity of energy, we consider different parts of a macroscopic system which are situated in different parts of ordinary space as subsystems. Strictly speaking, neighbouring parts are correlated with one another and therefore their entropy is not additive. We make, however, the same approximation as we did for the energy and neglect the correlation entropy. Then entropy becomes an extensive quantity too. This corresponds to the macroscopic standard description according to which entropy in a homogeneous state is proportional to volume.

> **Remark:** It should, however, be stressed that this approximation is not exactly the same as for energy because correlations between the parts are not the same as their energetic interaction. For instance by approach to a critical point of a phase transition long range correlations arise, whereas the interaction energy remains short range. It may be even that there are correlations in absence of any interaction. This is the case in the interaction free *Fermi* gas where the exclusion principle gives rise to correlations between the particles.

Internal energy. In addition to eq. (1.1), another identification is necessary to establish the correpondence between the statistical theory and phenomenology. This is the identification of the mean value of energy H with the so-called *internal energy* U of phenomenology:

$$U = \text{tr}\,(\rho\,\mathbf{H}). \tag{1.3}$$

The name "internal" is used because the energy of the macroscopic motion of the system as a whole is not included. U is a thermal quantity belonging only to the thermal state. Thus it includes only the internal molecular motion. That is to say, we consider systems which are in rest as a whole.

In phenomenology, for S in equilibrium, the relation

$$\frac{\partial S}{\partial U} = \frac{1}{T} \tag{1.4}$$

is satisfied, where T is the *absolute temperature*. The derivative in eq. (1.4) has to be taken for fixed values of all other extensities M^ν which are different from U. We shall choose $\mathbf{M}^0$ as $\mathbf{H}$. The corresponding λ_0 is designated by the letter β, according to a common tradition in statistical mechanics:

$$\beta = \frac{1}{kT}. \tag{1.5}$$

For the purpose of finding a correspondence between quantities of phenomenological thermodynamics and quantities occurring in the statistical theory, we adopt eq. (1.4) as definition of "absolute temperature" in statistics as well. We do not anticipate further properties of macroscopic temperature. It is rather the aim of the statistical theory to show that they are fulfilled indeed by the quantity T of statistics. The first of these properties is a particular consequence of the general theorem of equal intensity values of different systems in an absolute equilibrium:

> Systems which can exchange energy without restriction assume equal temperature in equilibrium with each other.

This is just the basis of the measuring prescription of temperature: The thermometer has to be brought into equilibrium with the system to be observed.

2.2.2 Work and Heat

Working parameters. Internal energy U which we identified with M^0 is a distinguish extensity occurring practically always in the set M of an equilibrium state. In the following we shall use a Roman letter like l as superscript of the other extensities instead of the Greek letter like v. If a Greek index runs from 0 to say r, we will use a Roman letter if it runs only from 1 to r. We have to distinguish two different classes of extensities. The first one represents variables M^l the change of which does not require a change of the material composition of the system but in general involves a change of its energy. This class comprises volume of a gas in a piston, surface of a droplet, other parameters describing the extension and shape of a deformable body, or magnetization of a magnet. We call this class *working parameters*. Extensities of the second class always require an increase or decrease of the matter of the system. Typical quantities belonging to this class are particle numbers of different species. As in the standard applications volume V is the most frequently occurring working parameter, we choose the symbol V^l for the whole class.

Of particular interest are state changes which occur so slowly that the system practically passes thermal states only. This condition is fulfilled if there is a time scale separation between the fast equilibrium to an instantaneous equilibrium and the slow change of the thermal state. If, however, this slow process occurs extremely slowly, then it can — at least in an idealization — be reversed in all steps. Such processes are called *quasistatic* or *reversible*. We can influence quasistatic changes of the working parameters by adequate macroscopic devices arbitrarily. We say, we can "control" these parameters.

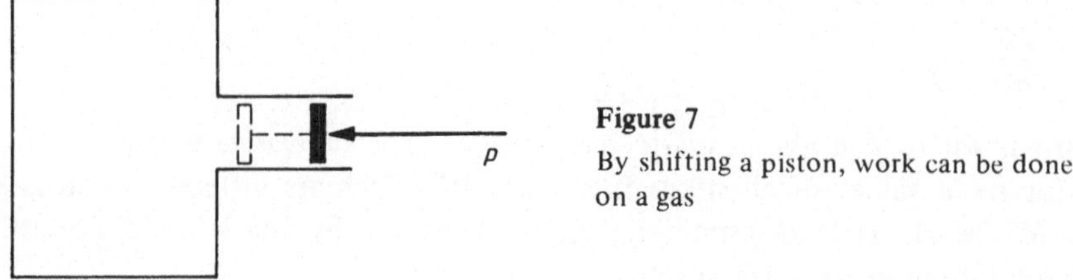

Figure 7

By shifting a piston, work can be done on a gas

Returning to the example of the *gas in a piston*, we assume that the only influence on the system is possible by a motion of the piston (compare Figure 7). In all other respects it shall be isolated from other systems. We shall call the energy change in this case *work*. If the change of the volume V is very small, say "infinitesimal" with respect to macroscopic dimensions and if it is moreover quasistatic, we can write the increase of energy of the system in the form

$$đW = -p\, dV. \tag{2.1}$$

We call it the work done "on" the system. By definition, the work is positive if it increases the energy of the system. This is the case when the volume decreases. The work is done by shifting the piston with the cross section say ω over a distance ds into the direction against the interior of the gas vessel. This yields

$$dV = -\omega\, ds. \tag{2.2}$$

The mechanical force of the gas against the piston is $p\,\omega$. This means that p is the *pressure* of the gas acting on the piston. The generalization to more working parameters yields

$$đW = -p_l\, dV^l, \tag{2.3}$$

where the equal sub- and superscript l is again a summation dummy. We designate the differential symbol here and in eq. (2.1) by a bar because it is not an exact (or "total") differential of a function of the state variables M^ν. We shall return to this point in detail in chapter 2.3.1.

We consider now a system the material composition of which remains unchanged. Then all occurring M^ν are working parameters V^l. We can write the *Gibbs* fundamental equation (1.2) in the form

$$dU = -p_l\, dV^l + T\, dS. \tag{2.4}$$

With eq. (2.4) we have put

$$\lambda_0 = \frac{1}{kT}, \qquad \lambda_l = \frac{p_l}{kT} \tag{2.5}$$

The quantities p_l are also called "intensities" and "thermal conjugates" of the corresponding extensities V^l. Because of the principle that each independent λ_ν is equal for systems in unrestricted equilibrium with one another, the same is true for T and p_l. These quantities, therefore, are homogeneous in a system in unrestricted equilibrium. In particular, pressure p then is acting equally on all parts of the walls of a vessel in which a gas or a fluid is enclosed. Moreover, the same pressure is acting on any interface between two parts of the interior of the vessel, from one part to the other one and vice versa.

> **Remark:** Special cases like the barometer pressure in the gravitational field or the pressure in a droplet with surface tension do not follow this rule because the conditions are not fulfilled This will be discussed with examples in chapter 3.2.4 and 2.3.7.

Because of eq. (2.4) the increase dU of internal energy consists of two parts. The first part is the work $đW$. The second part will be considered in the following.

Heat. The change dU of eq. (2.4) is different from dW if the system is influenced by the environment not only on account of changes of the working parameters. The difference represents an energy exchange with the environment which is not controlled by macroscopic devices in the same way as dW is. We call it *heat* transferred to the system:

$$dQ = T\,dS. \tag{2.6}$$

It should be stressed that it is heat transfered by a reversible change. In irreversible changes the uncontrolled part of transferred energy can be different from that of eq. (2.6). Nevertheless it is also called "heat".

The First Law of thermodynamics. Energy is a conserved quantity in any dynamics of the microstates in the sense that it cannot be generated or destroyed anywhere. This is of course also true for the mean value, the macroscopic internal energy. If an isolated system is divided into two subsystems with internal energies U_I and U_{II} respectively, in macroscopic thermodynamics the internal energy of the total system is

$$U = U_I + U_{II} \tag{2.7}$$

because interaction energy can be neglected, as explained in chapter 2.1.6. Due to the energy conservation, for any state change of the isolated system

$$dU_I + dU_{II} = 0. \tag{2.8}$$

Another consequence for the isolated system is

$$dW + dQ = 0. \tag{2.9}$$

> Work can be exchanged between systems, yet it can be produced or destroyed only by transformation of heat into work and vice versa.

This principle is called the *First Law* of thermodynamics.

Two systems in which no working parameters are changed can exchange energy only in form of heat. Heat then behaves like a conserved quantity. This was the reason why heat originally was observed as a particular measurable quantity. To measure it, so-called *calorimetric* processes of pure heat exchange were performed and the change of temperature connected with the process, was observed.

Irreversible changes. Now let us consider a finite change ΔV^l of the working parameters by a process performed too fast to be quasistatic. First let us speak of the special case that the intensities T, p_l of the environment, which is assumed to be very large, are not changed. Then the changes of the environment are quasistatic and the increase of its entropy is

$$\Delta S_e = -\frac{\Delta Q}{T} \tag{2.10}$$

if ΔQ is the heat transferred to the system from the environment. Without restriction of generality we can assume that the system and the environment form an isolated system together On account of the Second Law, the entropy of this compound system cannot decrease. If ΔS is the change of entropy of the system by the process, then

$$\Delta S_e + \Delta S \geqslant 0. \tag{2.11}$$

That is with heat ΔQ added to the system

$$\Delta S \geqslant \frac{\Delta Q}{T}. \tag{2.12}$$

When the process is finished, after a certain time the system assumes again the values T, p_l of the environment which it had before the process. Now we consider a process in which the dates of the environment vary in time. It is a succession of steps with small amounts $\mathrm{d}Q$ of heat transferred to the system. As for each step inequality (2.12) holds, we obtain for the whole increase of entropy of the system

$$\Delta S \geqslant \int \frac{\mathrm{d}Q}{T} \tag{2.13}$$

valid for an irreversible process.

2.2.3 Standard Distributions

The generalized canonical distribution associated with a given set of mean values M^ν is

$$\rho = \exp\left(\Psi - \lambda_\nu M^\nu\right). \tag{3.1}$$

As settled in chapter 2.2.1, $\mathbf{M}^0$ shall be energy $\mathbf{H}$ with the mean value U called "internal energy". The thermally conjugate λ_0 of U is β of eq. (1.5), that means $1/kT$ with abolute temperature T. Depending on which further quantities $\mathbf{M}^\nu$ enter into the distribution, we obtain different kinds of distributions ρ. The normalization constant Ψ in these different distributions has different names and designations by convention. It is connected with ρ by eq. (6.4) of chapter 2.1.6.

The canonical distribution. So is called the distribution which corresponds to the situation that the temperature is given as the only thermal variable. This is realized with a system in a so-called *heat bath* or *heat reservoir*, i.e. an environment with temperature T with which the system is in thermal equilibrium. T also can be registered with a thermometer brought into contact with the system directly. With given T internal energy is also determined. Therefore ρ can be interpreted to be the result of the unbiased guess to given mean value U of energy $\mathbf{H}$. The corresponding distribution is usually written in the form

$$\rho = \exp \beta (F - \mathbf{H}), \tag{3.2}$$

where

$$F = -\frac{1}{\beta} \ln \exp \operatorname{tr}(-\beta \mathbf{H}) \tag{3.3}$$

is called *Helmholtz free energy*. Entropy is

$$S = \frac{1}{T}(U - F).\tag{3.4}$$

The pressure ensemble. If temperature T and pressure p are given, we obtain another distribution. It belongs to a system the volume V of which is a random quantity. This is realized by a system with movable walls in an environment acting with pressure on the system, for instance on a piston filled with gas. The corresponding distribution

$$\rho = \exp \beta (G - \mathbf{H} - p\mathbf{V})\tag{3.5}$$

is called the *pressure ensemble*. G is called *Gibbs free energy* and satisfies

$$G = U + pV - TS = F + pV.\tag{3.6}$$

This distribution is representative also for the general case that other working parameters V^l are given as mean values, or that instead of these values the corresponding conjugate thermal variables p_l are given. In this case we can adopt eqs. (3.5, 3.6) if replacing pV with the sum $p_l V^l$. In the following we shall often take V as symbol for all working parameters V^l. Then we shall in future obtain more general relations if replacing p and V with p_l and V^l respectively. Then pV will be replaced with the sum $p_l V^l$. We shall discuss in particular, however, the following case of a working parameter different from volume.

The magnetic field ensemble. This distribution belongs to a magnetic material with magnetization $\vec{M}$ in a magnetic field $\vec{B}$. In future we shall mark vectors and pseudo-vectors in ordinary three-dimensional space with an arrow. To make things simple, we speak of a part of the material of unit volume. The work done on this piece by a change $d\vec{M}$ of the magnetization is

$$đW = \vec{B}\, d\vec{M}.\tag{3.7}$$

Any component of the three-dimensional pseudovektor $\vec{M}$ is a working parameter. Correspondingly we obtain the so-called *magnetic field ensemble*

$$\rho = \exp \beta (G - \mathbf{H} + \vec{B}\vec{\mathbf{M}}).\tag{3.8}$$

Here the *Gibbs* free energy is

$$G = U - \vec{B}\vec{M} - TS = F - \vec{B}\vec{M}.\tag{3.9}$$

The grand canonical ensemble. This is the distribution belonging to the case that the unbiased guess is based not only on given mean values of working parameters V but on particle numbers N^a of different species as well. This corresponds to a system which can exchange particles with the environment, or in which chemical reactions are possible:

$$\rho = \exp \beta (\Phi - \mathbf{H} - p\mathbf{V} + \mu_a \mathbf{N}^a).\tag{3.10}$$

The index a is a dummy running over all species of particles with varying particle numbers. The N^a are not working parameter as already explained. Therefore the intensities

μ_a are not mechanical "forces" like p. They are called *chemical potentials* of the different chemical species.

It should be stressed that in macroscopic physics not the particle numbers are directly measurable, but the numbers of *moles*. One mole of a chemical substance is the quantity of m grams if m is the chemically observable "molecular weight" of this substance. One mole of any substance comprises the same number of molecules independent of the weight. This number is called the *Loschmidt number L* and is of the order $6 \cdot 10^{23}$ as for instance experiments show wich will be discussed in chapter 5.1.3. The number of moles is just the number of particles divided by L. In case that N^a is not the particle number but the number of moles, μ_a is the chemical potential per mole and not per particle, i.e. by the factor L larger than the latter one.

The quantity

$$\Phi = U + pV - \mu_a N^a - TS = G - \mu_a N^a \tag{3.11}$$

has no special name. The *Gibbs* fundamental equation (1.2) can be written in the form

$$T \, dS = dU + p \, dV - \mu_a \, dN^a. \tag{3.12}$$

The chemical potentials. If the species a is allowed to move freely in the system, in equilibrium the chemical potential is spatically homogeneous in the interior of the system, like pressure and temperature. (This is true if no external force like gravitation or an electrical field is involved. These cases will be discussed in a later chapter.)

To better elucidate the role of the chemical potential, we consider a substance in a mixture which is able to diffuse through a wall between two vessels. Before reaching the equilibrium, the chemical potential may have two different values μ', μ'' in the two vessels. The vessels may form an isolated system together so that U and V are constant. The total particle number N of the substance is constant as well. Its time derivative satisfies

$$\dot{N} - \dot{N}' + \dot{N}'' = 0. \tag{3.13}$$

The total entropy never decreases. If U and V remain constant, we thus obtain due to the *Gibbs* fundamental equation (3.12):

$$T\dot{S} = - (\mu' - \mu'') \dot{N}' \geqslant 0. \tag{3.14}$$

In the average the particles move from higher to lower values of μ. The direction of this motion is analogous to that of particles in a mechanical potential if they start at rest. This is the reason for the name chemical "potential". The origin of both motions, however is of totally different nature. The diffusing particles are not driven by a mechanical force but by the statistical trend to assume macrostates which are realizable by more microstates and have higher entropy.

The equipartition theorem. This is a theorem valid only in classical mechanics. It says that each degree of freedom contributes with the same amount to the mean value of the kinetic energy. To prove it, we consider a classical mechanical system with f degrees of freedom. If the system is conservation and if q_r are independent configuration coordinates, the kinetic energy is always a quadratic form of the canonically conjugate momenta p_r:

$$H_{kin} = \frac{1}{2} \sum_{rs} p_r\, a_{rs}(q)\, p_s. \tag{3.15}$$

r and s run form 1 to f. In general, the matrix a_{rs} may depend on the set q of all q_l yet not on the set p. In the canonical distribution, the probability density of the momenta p with fixed q is a normal distribution

$$\hat{\rho}(p) = C(q) \exp\left[-\frac{\beta}{2} \sum_{rs} p_r\, a_{rs}(q)\, p_s \right]. \tag{3.16}$$

Referring to chapter 1.1.5 we see hat βa_{rs} is the reciprocal matrix of the correlation matrix

$$\gamma_{rs} = \langle p_r p_s \rangle. \tag{3.17}$$

Thus for any subscript r and any set q:

$$\sum_s \beta a_{rs} \langle p_r p_s \rangle = 1. \tag{3.18}$$

By summation over the f subscripts r and by averaging over all values of the set q,

$$\langle H_{kin} \rangle = \frac{1}{2} \sum_{rs} \langle a_{rs} \rangle \langle p_r p_s \rangle, \tag{3.19}$$

we obtain

$$\langle H_{kin} \rangle = f\, \frac{kT}{2}. \tag{3.20}$$

This so-called *equipartition theorem* states:

> Any degree of freedom of a conservative classical mechanical system contributes with the same amount $kT/2$ to the mean value of the kinetic energy in thermal equilibrium.

This theorem has several important consequences we shall discuss later. Here we are interested in the following one: The *Boltzmann* constant k occurring in β of course depends on the choice of units of U and T. Independent of the choice, however, is the ratio U to kT. In a macroscopic system the number f of degrees of freedom is of an order of about 10^{23}. This is the order of the considered ratio. We see that kT is extremely small in macroscopic units.

2.2.4 Extensies as Sharp Parameters

Up to now we have been describing a thermal state given by a set of extensities M^ν using a generalized canonical distribution in which all these quantities appeared as random quantities. Their mean values were identified with the macroscopic values. Often also another description is used in which some of the extensities are not random quantities but parameters of the distribution ρ. These parameters then have the macroscopic value of these extensities as a "sharp" not fluctuating value. For instance, the gas in a piston can equally be described by the pressure ensemble with fluctuating volume corresponding to a movable piston, or by a canonical distribution for fixed volume of the macrospically given value. On account of the macroscopic size of the system, the observation of the piston in equilibrium will fit both models because the relative fluctuations are extremely small.

We shall discuss the situation on a more general level. If we exclude such exceptional phenomena like phase transitions, we can say that the equilibrium fluctuations of an extensity M^{ν} are extremely small in a macroscopic system. Fluctuations and susceptibilities are connected with each other by eqs. (2.23, 2.27) of chapter 1.3.2:

$$\langle \Delta M^{\nu} \Delta M^{\sigma} \rangle = - \frac{\partial M^{\nu}}{\partial \lambda_{\sigma}} . \tag{4.1}$$

This gives for each M^{l}:

$$\frac{\langle (\Delta M^{l})^{2} \rangle}{\langle M^{l} \rangle^{2}} = - \frac{kT}{(M^{l})^{2}} \frac{\partial M^{l}}{\partial y_{l}} . \tag{4.2}$$

The square root of this quantity is a measure of the relative width of the fluctuations of the extensive quantity M^{l}. The multiplier kT on the right hand side of the last equation is extremely small compared with the other part of the product which is a macroscopic quantity. This gives rise for neglecting the fluctuations and for the replacement of the random quantity $\mathbf{M}^{l}$ in the distribution ρ by the sharp value M^{l}. This approximation is called the *saddle point method*.

If, using this method, we replace for instance the pressure ensemble by the canonical distribution, the volume appears as a parameter V in the *Hamiltonian* $\boldsymbol{H}$. Then F, U, S become dependent on V. We can introduce *pressure* p in the following way. *Helmholtz* free energy $F(T, V)$ is connected with *Gibbs* free energy G by

$$G = F + pV. \tag{4.3}$$

For the pressure ensemble

$$T\, \mathrm{d}S = \mathrm{d}U + p\, \mathrm{d}V. \tag{4.4}$$

With eq. (3.6)

$$G = U + pV - TS \tag{4.5}$$

eq. (4.4) obtains the form

$$\mathrm{d}G = V\, \mathrm{d}p - S\, \mathrm{d}T, \tag{4.6}$$

which in particular yields

$$\left(\frac{\partial G}{\partial p} \right)_{T} = V. \tag{4.7}$$

Here and in the following we shall use a notation conventional in thermodynamics. We add as subscript to the bracket of a partial differential quotient the symbols of the quantities which have to be kept constant during the differentiation. Due to eq. (4.7), eq. (4.3) becomes a *Legendre* transformation from the variables p to V, yielding

$$\left(\frac{\partial F}{\partial V}\right)_T = -p. \tag{4.8}$$

This equation now is used as the definition of pressure in the canonical distribution.

It is easy to generalize from V to an arbitrary extensity M^r in a generalized canonical distribution

$$\rho = \exp \beta (J - \mathbf{H} - y_l \mathbf{M}^l). \tag{4.9}$$

Let r be the last of the numbers

$$l = 1, 2, \dots , r. \tag{4.10}$$

The *Gibbs* fundamental equation

$$T \, \mathrm{d}S = \mathrm{d}U + y_l \, \mathrm{d}M^l \tag{4.11}$$

and

$$J = U + y_l M^l - TS \tag{4.12}$$

yield

$$\mathrm{d}J = M^l \, \mathrm{d}y_l - S \, \mathrm{d}T, \tag{4.13}$$

and in particular

$$\left(\frac{\partial J}{\partial y_l}\right)_T = M^l. \tag{4.14}$$

If we replace $\mathbf{M}^r$ with the sharp parameter $\mathbf{M}^r$, the distribution ρ is changed into

$$\rho' = \exp \beta (J' - \mathbf{H} - y_s \mathbf{M}^s), \tag{4.15}$$

where s runs from 1 to $r - 1$ only and where

$$J(y_1, \dots , y_r) = J'(y_1, \dots , y_{r-1}, M^r) + y_r M^r. \tag{4.16}$$

(r here is not a summation dummy!). This equation is the generalization of eq. (4.3). With eq. (4.14) it represents a *Legendre* transformation from the independent variable y_r to M^r and yields

$$\left(\frac{\partial J'}{\partial M^r}\right)_{y_s, T} = -y_r. \tag{4.17}$$

This equation shows how y_r is obtained by ρ'.

Generally we can say that the macroscopic relations eqs. (4.11, 4.12) are unchanged if we replace ρ with ρ'. These macroscopic relations, however, are the basis for a general scheme of thermodynamics.

The microcanonical distribution. This is the distribution in which also energy is described by a sharp parameter U. As explained already in chapter 2.1.3, to apply the principle of equal a priori probability, we have to consider a thin energy shell

$$U - \Delta U \leqslant H(\xi) \leqslant U \tag{4.18}$$

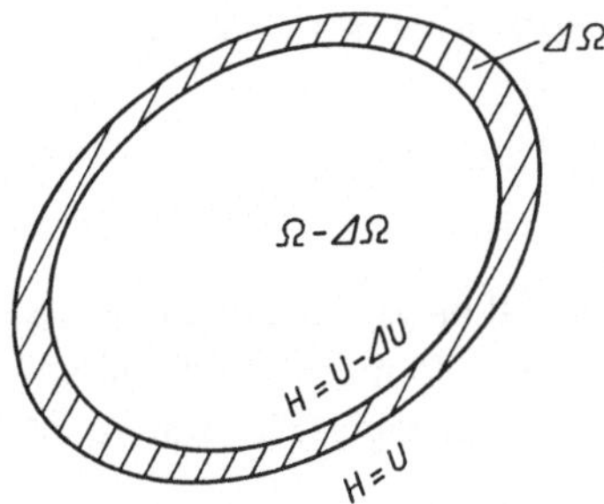

Figure 8

Energy shell in phase space

as allowed region in the Γ-space instead of an energy surface that has volume zero (see Figure 8). The unbiased guess now yields equiprobability in this region. If $\eta(\xi)$ is a function which has the value 1 in this region and zero on the outside, then we can write

$$\rho(\xi) = \frac{1}{\Delta\Omega} \, \eta(\xi), \tag{4.19}$$

where $\Delta\Omega$ is the volume of the shell. As we have been concerned with systems in thermal equilibrium so far, we can assume that the considered energy surface encloses a restricted volume in Γ-space. With introduction of the *Heavyside* step function $\Theta(x)$, defined on the real x-axis to be 1 for positive x and zero for negative x, we can write for the volume enclosed by the closed surface U in the form

$$\Omega = \int d^{2f}\xi \, \Theta(U - H(\xi)). \tag{4.20}$$

With the *Dirac* delta-function $\delta(x)$, which is the derivative of the step function $\Theta(x)$, we can write in the limit of vanishing ΔU

$$\rho(\xi) = \frac{1}{\omega} \, \delta(U - H(\xi)), \tag{4.21}$$

where

$$\omega = \int d^{2f}\xi \, \delta(U - H(\xi)) = \frac{d\Omega}{dU}. \tag{4.22}$$

This $\rho(\xi)$ is the so-called *microcanonical distribution* already introduced in chapter 2.1.3. The entropy that belongs to this is, up to an additive constant, which we put zero by definition,

$$S = k \ln \Delta\Omega. \tag{4.23}$$

Now we assume that the system is very large. This means that the dimension $2f$ of the Γ-space is say roughly of an order by 10^{23}. Energy U can be considered to be only one of the degrees of freedom. Taking into account solely the orders of magnitude, we obtain

$$\Omega \approx U^{2f}. \tag{4.24}$$

Any increase δU of U which is small compared with U is associated with an increase $\delta \Omega$ of Ω for which

$$\frac{\delta \Omega}{\Omega} \approx 2f \frac{\delta U}{U}. \tag{4.25}$$

Even if δU is small on a macroscopic scale, $\delta \Omega$ will still be very much larger than Ω. This is a puzzling fact in the geometry of high dimensions. It means that the volume Ω is located practically only on the surface. Therefore we can replace $\Delta \Omega$ in eq. (4.23) with Ω and obtain

$$S = k \ln \Omega. \tag{4.26}$$

Temperature T is defined by eq. (1.4). This yields

$$\frac{\partial S}{\partial U} = \frac{1}{T} = k \frac{\omega}{\Omega}. \tag{4.27}$$

If working parameters V^l occur as sharp parameters in the *Hamiltonian* $H(\xi)$, we can obtain the adjoint intensities p_l by the relation

$$\left(\frac{\partial S}{\partial V^l}\right)_U = \frac{p_l}{T} \tag{4.28}$$

in accordance with the *Gibbs* fundamental equation. So we find

$$\frac{p_l}{T} = \frac{k}{\Omega} \left(\frac{\partial \Omega}{\partial V^l}\right)_U = -\frac{k}{\Omega} \int d^{2f} \xi \, \Theta'(U - H) \frac{\partial H}{\partial V^l}, \tag{4.29}$$

where the derivative $\Theta(x)$ is $\delta(x)$. Thus, with eqs. (4.21, 4.27)

$$p_l = -\left\langle \frac{\partial H}{\partial V^l} \right\rangle. \tag{4.30}$$

This is the mean value of a generalized mechanical force already defined in the micro-states.

It is worthwhile to compare the work

$$dW = \left\langle \frac{\partial H}{\partial V^l} \right\rangle dV^l \tag{4.31}$$

with

$$dU = \frac{\partial}{\partial V^l} \langle H \rangle \, dV^l. \tag{4.32}$$

In the first expression the energy shell and thus ρ and U are kept constant. In the last expression the energy shell is shifted by the change dV^l. The difference between dU and $đW$ is the heat $đQ$ transfered to the system.

2.2.5   The *Carnot* Cycle

As already mentioned, in phenomenological thermodynamics the distinction between work and heat appears already in the starting points of the theory, whereas in the statistical theory it appears in a relative late state of the theory in a more secondary way. Work is an energy transfer controlled by macroscopic mechanical devices, and heat is the remaining energy transfer not controlled in the same way. In phenomenology it is a central question to which degree heat can be transformed into work. This question was decisive in the development of thermodynamics. *S. Carnot* was the first to recognize that the transformation is always connected with the transition of heat from a warmer to a colder system. *Carnot* therefore studied a device that returns to its initial state by a "cycle" process.

In this cycle which is called a *Carnot cycle* the device, called a *Carnot engine*, withdraws a certain amount Q_2 of heat from a heat bath of temperature T_2. The *heat bath* or *heat reservoir* is a large system of practically unchanged temperature during the exchange of heat with the engine. The engine transforms a part of the heat Q_2 into work and delivers the rest of the heat to a reservoir of lower temperature T_1 (schematically sketched in Figure 9). As we shall call all energy positive if given to the engine, and negative if withdrawn from the engine, the heat delivered from the engine to the colder reservoir will be called $-Q_1$ and the work produced by the engine and delivered elsewhere $-W$.

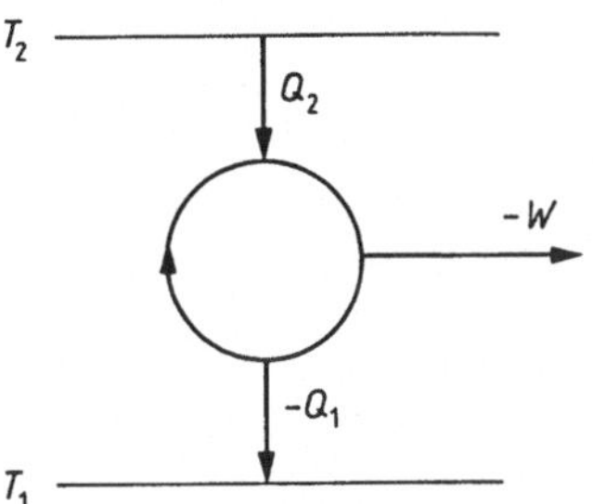

Figure 9

The scheme of a heat engine working between two heat reservoirs of temperature T_2 and T_1, respectively. By a cycle process the heat Q_2 is delivered to the engine by the hotter reservoir, the heat $-Q_1$ is withdrawn from the engine and given to the colder reservoir, and the work $-W$ is produced by the engine.

We shall discuss the *Carnot* cycle first using the results we have already obtained with the statistical theory. In chapter 2.3.1 we shall discuss it once more but in the framework of the basic principles of phenomenology. Then we shall be led to the concept of absolute temperature and phenomenological entropy.

The engine finally returns into the initial state and thus does not change its energy $U.$ — We obtain

$$W + Q_1 + Q_2 = 0. \tag{5.1}$$

Also the entropy of the engine, being a unique function of the thermal state, will assume the initial value again after the cycle is closed. The heat reservoirs do not change their temperature and are so large that their change of states is practically quasistatic. Then the increase of the sum of entropy of the two reservoirs is

$$\Delta S = -\frac{Q_1}{T_1} - \frac{Q_2}{T_2}. \tag{5.2}$$

This is also the entropy increase of the whole system composed of engine and reservoirs. This compound system can be supposed to be isolated. On account of the Second Law, ΔS is never negative:

$$\frac{Q_1}{T_1} + \frac{Q_2}{T_2} \leqslant 0. \tag{5.3}$$

The ratio of produced work to the heat Q_2 withdrawn from the warmer reservoir is called the *efficiency* of the cycle:

$$\eta = -\frac{W}{Q_2} = 1 + \frac{Q_1}{Q_2}. \tag{5.4}$$

We compare this *forward cycle* in which

$$Q_2 > 0, \qquad W < 0, \tag{5.5}$$

with a cycle passing in the opposite direction, called a *backward cycle* in which

$$Q_2 < 0, \qquad W > 0. \tag{5.6}$$

Whereas in the forward cycle the engine produces work, in the backward cycle it pumps heat from the colder to the warmer reservoir by consumption of work energy, delivered from elsewhere. Then it is a *heat pump*.

For the forward cycle it holds that

$$\frac{Q_1}{Q_2} + \frac{T_1}{T_2} \leqslant 0. \tag{5.7}$$

For the backward cycle the inequality sign has to be inverted. We call the efficiency η of eq. (5.4) in the forward cycle η_+, and in the backward cycle η_-, and obtain

$$\eta_+ \leqslant 1 - \frac{T_1}{T_2}, \qquad \eta_- \geqslant 1 - \frac{T_1}{T_2}. \tag{5.8}$$

Now we consider a quasistatic and thus *reversible cycle*. For this one let η be called η_r. It has to satisfy both relations, (5.8) which means that

$$\eta_r = 1 - \frac{T_1}{T_2}. \tag{5.9}$$

It is the maximum η_+ and belongs to the optimal gain of work. We emphasize as fundamental result:

> It is impossible to transform heat totally into work without another change elsewhere.

2.3 The Phenomenological Framework

In the preceding chapters we deduced results which form the fundamentals of macroscopic thermodynamics on the basis of the statistical theory. In the historical development of thermodynamics, these fundamentals were obtained earlier than their foundation through a statistical theory. They were laws induced and conformed by experience on a macroscopic level. We call thermodynamics in this state "phenomenology". For a long time, the special means of description and the special measuring methods used in this field of physics were not put in relation to molecular motion or other elements of microscopic dynamics. In phenomenology specific methods of logical dedection were developed. The rational structure of this theory is still attractive today for extended analytical research, in particular in mathematical and logical respects. Different axiomatic systems of thermodynamics were developed after thermodynamics already had been a well established part of physics. They have gained the attention primarily of physical mathematicians and have the purpose to analyse the impressive logical structure of phenomenological thermodynamics. For the applied physicist and chemist, however, it is in real cases often easier, for instance, to decide whether a system will assume a thermal state than to define the conditions for a thermal state on the basis of an axiomatic system which avoids any logical abundance. The requirement for logically independent axioms is different from the design favored here, namely to start with laws which are as transparent with respect to their consequences as possible. Within the phenomenological framework, we shall take the existence of thermal states as a given fact. They are special states of a large class of pyhsical systems and can be described by relative few thermal variables.

In this subsection we shall represent the phenomenology as deduced from its own basic laws, not only because the strikingly closed logical structure. Another reason is that phenomenology can be applied successfully in a wide field of experience without the knowledge of statistics. The basic laws in the phenomenological framework have the character of postulates. Nevertheless, in this subsection we shall sometimes point to the statistical theory of which they result from and which was represented in the preceding section 2.2. With respect to the afore-mentioned logical structure of the phenomenological theory, we shall purposely introduce some concepts once more and independently of the preceding chapters, even if they already there occurred in a different context.

2.3.1 The Basic Laws of Thermodynamics

The postulates of the phenomenological theory which will be listed in this chapter are called "Basic Laws" of thermodynamics. Historically, the First and Second Law had been already formulated and discussed with respect to their consequences when a further postulate was formulated relatively late by some authors as a Basic Law of comparable rank. In the logical structure of the theory, however, this law has to be presumed prior to the others. Therefore it is called

> **The Zeroth Law.** Two systems which are in a thermal equilibrium with a third one are in mutual equilibrium with each other.

This law has already been formulated in chapter 2.1.6 as result of the statistical theory, but not like here as an independent postulate.

The First Law. This is the principle of energy conservation in thermodynamics. It states that work and heat are different forms of energy. Each one can be transformed into the other. The sum of both, however, is a quantity, the so-called "internal energy" that can never be produced or annihilated. This was the outstanding detection of *J. Robert Mayer* around the year 1843. As work and heat were originally considered to be different quantities corresponding to different measuring methods, they belonged to different definitions of units. The unit of heat was defined as one *calorie*. This is the amount of heat which raises the temperature of one gram of water by one degree centigrade. It is equivalent to 4.19 Joules, the so-called *mechanical heat equivalent*.

The conservation law of energy was already known in mechanics and later confirmed in electrodynamics. Today we know that it is a general consequence of time-shift invariance of any dynamics. In thermodynamics, not only the conservation of energy was the essential discovery but also the knowledge that heat is a form of energy.

If Q is an amount of heat transferred to a system and W is the work done on the system, then

$$\Delta U = Q + W \tag{1.1}$$

is the energy added to the system. The conservation of energy now means that if two systems I and II form an isolated compound system together, then for any change of their states

$$\Delta U_{\mathrm{I}} + \Delta U_{\mathrm{II}} = 0. \tag{1.2}$$

Energy can not arise or vanish anywhere. It can only be exchanged between the two systems by being shifted from one system to the other. A system may be led by a process from a state 1 into a state 2 and then be led back by another process into state 1 again. In Figure 10 these processes are represented by two different paths in a state diagram. The amount of energy ΔU which was withdrawn from the system by the first process must be given back by the second. The work W and heat Q transferred to the system by the first process will in general be different from work W' and heat Q' transferred from the system by the second process. But any process leading from 1 to 2

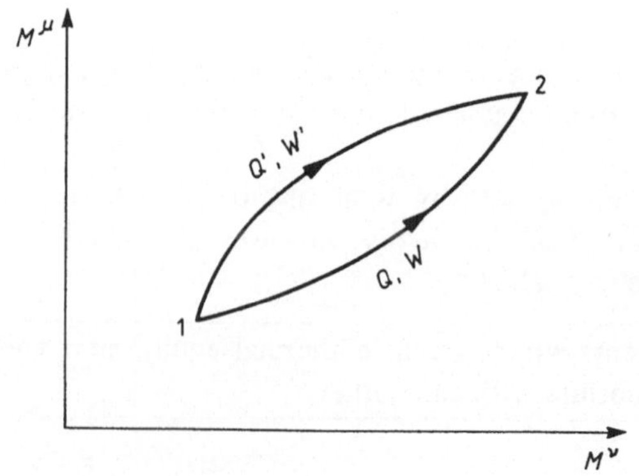

Figure 10

Illustration of the First Law of thermodynamics. For two different ways leading from a thermal state 1 to the same state 2 the needed heat and work may be different; but their sum, the needed energy, remains the same.

needs the same amount ΔU of energy which thus is uniquely associated with the state 2 if state 1 is given, unlike the work and heat which depend on the path, that means on the process. If we hold 1 fixed, ΔU becomes a unique function of state 2. We can choose a value U_1 arbitrarily and define a function U_2 of any state 2 by

$$\Delta U = U_2 - U_1. \tag{1.2}$$

Thus we can say, that to each thermal state of the system belongs an *internal energy U*. It is a so-called *state function*, i.e. uniquely associated with each thermal state. For infinitesimal state changes we write the increase of U in the form

$$dU = đW + đQ, \tag{1.3}$$

where $đW$ is the work and $đQ$ the heat transferred to the system. The latter two quantities are not unique but dependent on the process. To indicate this, we use the bar in the differential symbol, as will be explained in more detail in the following.

The most general form of work done by a quasistatic change is

$$đW = -p_l\,dV^l. \tag{1.4}$$

Both in the phenomenology and in the statistical theory, the "working" parameters V^l are defined as occurring in this general expression for work. Together with temperature they form a complete set of thermal variables. This means that this set defines the thermal state (in most cases) uniquely. As we did in the statistical theory, we shall often abbreviate the set of all working parameters V^l by the letter V and adopt a corresponding symbol p for the set of the thermal conjugates p_l.

The formal consequence of the First Law is that the state function *internal energy $U(T, V)$* exists, which up to an additive integration constant is uniquely defined with the exact (total) differential

$$dU = \left(\frac{\partial U}{\partial T}\right)_V dT + \left(\frac{\partial U}{\partial V^l}\right)_T dV^l. \tag{1.5}$$

In contrast to this, $đW$ in eq. (1.4) and

$$đQ = \left(\frac{\partial U}{\partial T}\right)_V dT + \left[\left(\frac{\partial U}{\partial V^l}\right)_T + p_l\right] dV^l \tag{1.6}$$

are mere *"Pfaff differential forms"*, and in general not integrable. That is, in general there does not exist a function of the state variables T, V the differential of which is $đW$ or $đQ$. This is the reason why we marked the differential symbol by a bar. Physically this means, that there is no "content" of heat or work energy as uniquely associated with the thermal state. There is, however, the "content" of internal energy U.

In a cycle process in which the system returns into the initial state, no energy can be gained or lost. An engine working periodically which produces energy was an old dream of mankind and was called a *perpetuo mobile*. We can formulate the First Law in form of

the principle of the impossibility of a perpetuo mobile.

The Second Law. This law can be formulated in different ways the equivalence of which is not so obvious. We start with the formulation obtained at the end of chapter 2.2.5 as result of the statistical theory:

> (1) It is impossible to transform heat entirely into work without other changes elsewhere.

This law now enters into the phenomenological theory as a postulate without any other proof besides the verification by its consequences.

A periodically working engine withdrawing heat from a reservoir and transforming the heat totally into work would not contradict the First Law; but this engine could not be in accordance with the Second Law. This hypothetical engine is called a *perpetuo mobile of second kind*. Equivalent to (1) is the formulation:

> (2) A perpetuo mobile of second kind can never exist.

We already considered the *Carnot* engine in chapter 2.2.5 using the concept of entropy of the statistical theory. We now consider this engine without any appeal to the earlier results, only using the following definition: A *Carnot* engine works in a cycle process between two heat reservoirs; it withdraws the heat Q_2 from the warmer reservoir, produces work $-W$ and delivers the heat $-Q_1$ to the colder reservoir.

The ratio

$$\eta = -\frac{W}{Q_2} \tag{1.7}$$

is called the *efficiency* of the engine. As already settled in chapter 2.2.5, the signs are defined to be positive for all quantities transferred to the system, negative if withdrawn from the system.

The First Lasw says that U of the engine is unchanged after it returned into the initial state. This yields:

$$Q_1 + Q_2 + W = 0, \tag{1.8}$$

$$\eta = 1 + \frac{Q_1}{Q_2} . \tag{1.9}$$

If W is positive, we again call the process a "forward" cycle and designate η by η_+. If W is negative, the process is called a "backward" cycle, the engine works as heat pump, and η is designated by η_-. The Second Law requires $\eta_+ < 1$. In the case $\eta_+ = 1$ which, however, is not permissible, we could be able to transform Q_2 totally into work. In case $\eta_+ > 1$ we could be able to transform not only Q_2 but beyond this even a part of $-Q_1$ into work. This is not possible and we can obtain a further formulation of the Second Law:

> (3) Heat cannot go from a colder to a warmer body without any change elsewhere.

If this were not true, we could use the two bodies as heat reservoirs of a *Carnot* engine.

In contradiction to the Second Law, the heat delivered by the engine to the colder body could go back to the warmer body. We should have obtained a perpetuo mobile of second kind.

The backward efficiency η_- of a heat pump can never be smaller than any forward efficiency. Otherwise the pump could bring more heat to the warmer reservoir than was withdrawn from by an engine which had produced the same work as the pump needs. This means, that heat would have gone from cold to warm without a change elsewhere. This would contradict (3). Thus we obtain

$$\eta_+ \leqslant \eta_-. \tag{1.10}$$

A reversible *Carnot* cycle is an idealized process with quasistatic changes only. It can be passed in forward and in backward direction. Its efficiency η_r is η_+ and η_- as well. Therefore η_r has to be maximum η_+ and minimum η_- simultaneously for all engines working between the same reservoirs. Thus η_r is equal for all reversible cycles. We can understand the last statement also by the following argument. If two reversible engines had different efficiencies, we could use the engine with the smaller one as heat pump to withdraw again the heat delivered to the colder reservoir by the other engine that acted forward. So finally the colder reservoir would remain unchanged; but heat withdrawn from the warmer reservoir would be transformed into work without any change elsewhere, in contradiction to (1). We obtain a further formulation of the Second Law:

> (4) All reversible *Carnot* cycles have the same efficiency if acting between the same reservoirs. It is the maximum efficiency of all forward cycles.

The following consequences of the law in from (3) are obvious:

> (5) Heat conduction is an irreversible process.
> (6) Heat production by friction is an irreversible process.

Irreversibility here is understood in the rigorous sense that not only the time-reversed process is impossible in which the system passes the same states like in a backward running moving picture. Rather it means that, in no way, it is possible to restore the initial state without any change elsewhere.

It should be stressed that all these statements from (1) to (6) are equivalent formulations of the Second Law. We can deduce all the others from any one. To give an example, if heat production by friction was reversible, we could transform heat into work wherever we wish. Or, if heat conduction which goes only from warm to cold were reversible, we could bring back the heat delivered to the colder reservoir by a forward *Carnot* engine and should have obtained a perpetuo mobile of second kind.

Not only (5) and (6) are forms of the Second Law. We can replace them by the existence of any irreversible process in which heat and work are involved. We can understand the Second Law as the law of irreversibility of certain processes.

Deductions of the kind presented here are characteristic for the specific logical methods in phenomenological thermodynamics we mentioned in the introductory part 2.3.

2.3.2   Definition of Absolute Temperature

The Second Law allows a definition of absolute temperature in the phenomenological framework, as will be shown in this chapter. The definition is independent of a thermometric substance. Therefore absolute temperature is a distinguished quantity.

Heat reservoirs of the same temperature can mutually exchange heat reversibly without work. Correspondingly no other property of the reservoirs can determine the reversible efficiency η_r than the temperature of the reservoirs used in the cycle. If the latter were not the case, we could use different kinds of reservoirs with the same temperature to construct a perpetuo mobile of second kind by using one reservoir in a forward, the other one in a backward cycle.

We start with an arbitrary temperature scale ϑ say defined by a special thermometric substance like e.g. mercury. We only suppose that the scale allows us to discriminate ϑ_1 to be colder than ϑ_2 by the inequality $\vartheta_1 < \vartheta_2$. The reversible *Carnot* efficiency

$$\eta_r = 1 - f(\vartheta_1, \vartheta_2) \tag{2.1}$$

is a universal function of ϑ_1, ϑ_2 and with it

$$f(\vartheta_1, \vartheta_2) = -\frac{Q_1}{Q_2} . \tag{2.2}$$

Now let us consider three heat reservoirs with temperatures

$$\vartheta_1 < \vartheta_2 < \vartheta_3 . \tag{2.3}$$

Two reversible *Carnot* engines M and M' acting between ϑ_1, ϑ_2 and ϑ_2, ϑ_3, respectively, can together form a compound engine acting between ϑ_1, ϑ_3 if the heat Q_2 withdrawn by engine M from the reservoir ϑ_2 is equal to the heat $-Q_2'$ which this reservoir received from the engine M'. Thus reservoir ϑ_2 finally remains unchanged (see Figure 11). This yields

$$f(\vartheta_1, \vartheta_3) = -\frac{Q_1}{Q_3'} = \left(-\frac{Q_1}{Q_2}\right) \left(-\frac{Q_2'}{Q_3'}\right). \tag{2.4}$$

The heat Q_3' withdrawn by machine M' from the reservoir ϑ_3 is also the heat withdrawn by the compound machine. Eq. (2.4) yields the functional equation

$$\ln f(\vartheta_1, \vartheta_3) = \ln f(\vartheta_1, \vartheta_2) + \ln f(\vartheta_2, \vartheta_3). \tag{2.5}$$

First this states that the expression on the left hand side is additive with respect to ϑ_1, ϑ_3. Consequently this is also true for the two terms on the right hand side with respect to their arguments. The therefore additive terms in ϑ_2 have to cancel each other. This leads to

$$\ln f(\vartheta_1, \vartheta_3) = a(\vartheta_1) - a(\vartheta_3). \tag{2.6}$$

If the designation ϑ_3 is replaced with ϑ_2, eq. (2.6) means

$$f(\vartheta_1, \vartheta_2) = T(\vartheta_1)/T(\vartheta_2) \tag{2.7}$$

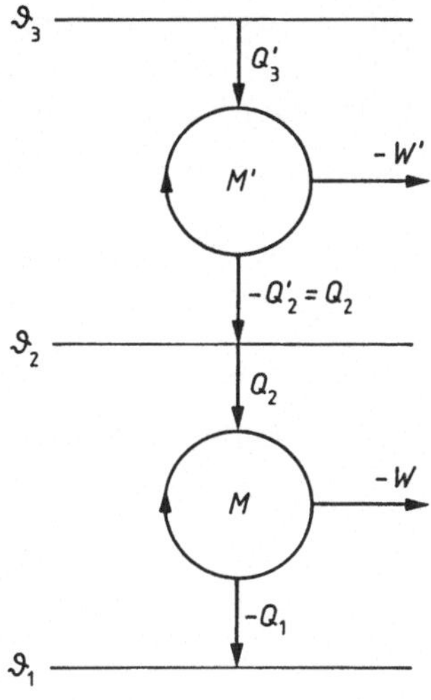

Figure 11

Illustration of the definition of absolute temperature. Two *Carnot* engines M' and M are coupled to a compound engine.

where $T(\vartheta)$ is a universal function $T(\vartheta)$. We call this the "absolute temperature". It is defined up to a factor dependent on the choice of the unit. The conventional choice is to settle the difference between the boiling and freezing point of water under standard conditions to be $100°$. With the introduction of absolute temperature we obtain for the reversible efficiency

$$\eta_r = 1 - \frac{T_1}{T_2} \,. \tag{2.8}$$

This relation opens many ways to measure T by adequate processes. Comparison with eq. (5.9) of chapter 2.2.5 shows that T is the same as in the generalized canonical distribution in the framework of the statistical theory.

2.3.3   Entropy in Phenomenology

The integrating factor of heat. The concept of entropy will be introduced in the phenomenological theory in a way considerably different from that we demonstrated in the statistical theory. To develop the phenomenological concept, we start from the Second Law in the form that a reversible *Carnot* cycle has the efficiency

$$\eta_r = 1 + \frac{Q_1}{Q_2} = 1 - \frac{T_1}{T_2} \,. \tag{3.1}$$

This yields

$$\frac{Q_1}{T_1} + \frac{Q_2}{T_2} = 0. \tag{3.2}$$

The reservoirs used in the cycle are assumed to be large compared with the *Carnot* machine itself. For the study of the exchange of heat between two systems we discuss the following situation. We may consider two arbitrary macroscopic systems 1 and 2 in thermal states as large reservoirs if the amounts of heat exchanged with a cycling machine acting between them are small enough, say $đQ_1, đQ_2$. For the following it is convenient to change the sign in the notation and in the future to designate the heat added to the system 1 with temperature T_1, and added to the system 2 with temperature T_2 respectively, by $đQ_1, đQ_2$. Then we can say, that for any reversible exchange of heat between such two systems always

$$\frac{đQ_1}{T_1} + \frac{đQ_2}{T_2} = 0. \tag{3.3}$$

In general such a reversible exchange requires that a further working system takes part. With the notation

$$đS_i = \frac{đQ_i}{T_i} \tag{3.4}$$

we can write

$$đS_1 + đS_2 = 0. \tag{3.5}$$

This equation is analogous to the conservation law of energy

$$dU_1 + dU_2 = 0 \tag{3.6}$$

for two systems which form an isolated compound system together. The consequence of this equation was that, up to an additive constant, U is a unique function of the thermal state, a "state function". It is not necessary for the validity of eq. (3.6) that the two systems form an energetically isolated system together. This only has to be isolated with respect to heat exchange, but in addition the change of state has to be reversible.

We can draw the analogous conclusion from eq. (3.5) as we did from energy conservation in chapter 2.3.1 leading to eq. (1.2):

> There is a state function S which behaves in reversible processes like a conserved quantity.

This function is called *entropy*. Its exact differential is

$$dS = \frac{đQ}{T}\,, \tag{3.7}$$

where $đQ$ is the heat added reversibly to the system. Now we can drop the bar in the differential symbol dS which we used as a precaution in eqs. (3.4, 3.5). The function S, like internal energy U, is defined up to an additive integration constant because only the differentials of these quantities were introduced by physical arguments. In mathematical terms, $1/T$ is an "integrating factor" which transforms the *Pfaff* differential form $đQ$ to an exact differential

$$dS = \frac{1}{T}(dU + p_l\,dV^l). \tag{3.8}$$

The existence of such an integrating factor can be considered as the content of the Second Law.

We now compare a reversible change for which

$$dU = T\,dS + đW \tag{3.9}$$

with an arbitrary state change, possibly also an irreversible one. Work $đW'$ and heat dQ' shall result in the same value of energy increase:

$$dU = đQ' + đW'. \tag{3.10}$$

The work $-đW$ produced by the reversible process cannot be smaller than the work $-đW'$ produced anyhow:

$$-đW + đW' = T\,dS - đQ' \geqslant 0. \tag{3.11}$$

So we obtain the inequality

$$dS \geqslant \frac{đQ'}{T} \tag{3.12}$$

for any stage change. A state change is called *adiabatic* if no heat exchange with another system occurs. For an adiabatic change the inequality

$$dS \geqslant 0 \tag{3.13}$$

holds.

> In an adiabaticly closed system entropy can never decrease with evolution of time.

A particular consequence of inequality (3.13) is:

> In an isolated system entropy never decreases.

This is a further formulation of the Second Law. For reversible changes in particular we obtain the result:

> In reversible adiabatic processes entropy remains unchanged.

Chemical potentials. So far we have defined entropy only for system which are materially closed. If we, however allow exchange of matter with other systems, we have to extend eq. (3.8) to

$$T\,dS = dU + p_l\,dV^l - \mu_a\,dN^a, \tag{3.14}$$

the so-called *Gibbs fundamental equation*. N^a are the particle numbers of the chemical components of the system. The extension of the entropy concept to eq. (3.14) is based on the introduction of the chemical potentials

$$\mu_a = -T\left(\frac{\partial S}{\partial N^a}\right)_{UV} \tag{3.15}$$

as thermal variables by postulate. That is to say, we can postulate the existence of the function entropy $S(T, V, N)$ depending on N^a for open systems as well and define the chemical potentials by eq. (3.15).

Instead of particle numbers the quantities N^a can be defined to the mole numbers. Then μ_a is the "molar" chemical potential and larger than the "molecular" chemical potential by the factor of the *Loschmidt* number L.

We emphasize that, in contrast to the phenomenological theory, in the statistical theory no new postulate was needed for the inclusion of materially open system when defining entropy.

In the following we shall often use the common designation M^l for all extensities, for working parameters and particle numbers as well. Then we shall write the *Gibbs* fundamental equation in the form

$$T\, dS = dU + y_l\, dM^l \tag{3.16}$$

where the letter y stands for p and $-\mu$, likewise.

*Energetic and Entropic Scheme. There are two different ways to interpret the *Gibbs* fundamental equation, depending on whether energy or entropy is regarded as the more fundamental quantity. By some authros these two interpretations are called the "energetic" or "entropic" scheme, respectively[8]). The two schemes are not contradictory and do not lead to different results. They suggest two different ways of deducing macroscopic phenomenology from experience. As today we seek the basis of thermodynamics in the statistical theory, the discrimination between the two systems has rather didactical significance than a fundamental one.

The *entropic scheme* interpretes the *Gibbs* fundamental equation in the form

$$dS = Y_\nu\, dM^\nu \tag{3.17}$$

which is identic with eq. (2.16) of chapter 1.3.2, i.e.

$$dI = -\lambda_\nu\, dM^\nu \tag{3.18}$$

with

$$Y_\nu = k\,\lambda_\nu. \tag{3.19}$$

The *energetic scheme*, however, is based on the conception that energy were the more fundamental quantity than entropy, because energy is the most important conserved quantity in physics. As already mentioned, energy is the generator of time shift in general dynamics. Time shift invariance of dynamical laws causes conservation of energy of the corresponding physical system. The preference of energy, however, is rather arbitrary. After all, entropy is a stochastic measure already defined in general statistics and is not restricted to physics.

In the entropic scheme, energy is one of the extensities M, say

$$U = M^0, \qquad Y_0 = \frac{1}{T}. \tag{3.20}$$

---

[8]) *P. Salamon, J. Nulton* and *E. Ihrig*, J. Chem. Physics **80**, 436 (1984).

In the energetic scheme, however, entropy S is put on a par with the remaining extensities M. Correspondingly we call it M^{r+1}. The *Gibbs* fundamental equation then obtains the form

$$\mathrm{d}U = y_l\,\mathrm{d}M^l, \qquad (l = 1, \dots, r+1). \tag{3.21}$$

y_l are the thermal conjugates to each M^l. Thus in this scheme

$$y_{r+1} = -T \tag{3.22}$$

is the thermal conjugate of entropy. Both schemes are connected by

$$y_\nu = TY_\nu. \tag{3.23}$$

This discrimination of these two schemes is not decisive for experimental results. Energy and entropy enter into the statistical theory in very different ways. Energy is defined already in the microstates; entropy, however, belongs merely to probability distributions over microstates independently of their individual properties. The two different interpretations have some significance for the question from which basic experience the macroscopic theory is deduced. The interpretation in terms of "heat" dominated the historical development. This corresponds to favoring calorimetric processes without work. Then heat is the only energy exchange and behaves like a conserved quantity. Therefore the way from this basis led to the energetic scheme. The entropic scheme, however, is primarily based on experience obtained by transformation of heat into work. This way of deducing thermodynamics is less popular. (It is worked out in a systematic approach in particular in a book by *G. Job* 1972[9]). In this book a new phenomenological concept of "heat" is used different from the traditional one. It leads very directly from experience to the concept of entropy.)

Without denying some advantages of the entropic scheme (as discussed for instance in references[9] and [10]), we shall preferably use the energetic scheme with intensities y like pressure and chemical potentials, to be in congruence with the conventional literature.

2.3.4 Thermodynamic Potentials

The three basic laws of thermodynamics — the Zeroth, First, and Second Law — can be comprised in the statement that the following quantities exist: thermal variables determining the thermal state, and moreover internal energy U, and entropy S as functions of the thermal variables. It is possible to change from one set of independent variables describing the thermal state to another one, say from the extensities to the intensities or to an independent mixture of both. Correspondingly it is adequate to change from entropy to other functions of thermal variables to obtain general relations, as will be explained in this chapter. These functions are called *thermodynamic potentials*.

---

[9] *G. Job*, Neudarstellung der Wärmelehre. Die Entropie als Wärme (Akademische Verlagsgesellschaft, Frankurt/M. 1972).

[10] *J. W. Gibbs*, "Collected Works", Vol. 1 (Yale Univ., Hew Haven 1948).

Their use is a powerful tool in thermodynamics. The different kinds of these potentials belong to the different conditions the physical system is subject to.

To elucidate the universal scheme, we begin with eq. (3.18) in the form

$$\frac{\partial I}{\partial M^\nu} = -\lambda_\nu. \tag{4.1}$$

If I, or equivalently, if entropy S is given as function of the extensities M^ν, we can gain the whole set of the intensities λ_ν by this equation. This is a situation similar to mechanics where a mechanical potential $\varphi(x)$ as function of all configuration coordinates x_k makes it possible to gain the components F_k of the forces by mere differentiation:

$$\frac{\partial \varphi}{\partial x_k} = -F_k. \tag{4.2}$$

On account of this analogy, $I(M)$ is called a thermodynamic "potential". By the *Legendre* transformation

$$\Psi(\lambda) = I(M) + \lambda_\nu M^\nu \tag{4.3}$$

we obtain

$$\frac{\partial \Psi}{\partial \lambda_\nu} = M^\nu. \tag{4.4}$$

This is the reason for calling $\Psi(\lambda)$ a thermodynamic potential as well. It has to be emphasized that I and Ψ are such potentials only if taken as function of the adequate variables M or λ respectively. These then are called the *natural* variables of the corresponding potentials. The derivative of such a potential with respect to one natural variable in eq. (3.17) has to be performed by keeping the other natural variables constant. The corresponding is true for eq. (4.4), and will be valid for further thermodynamic potentials which can be introduced by *Legendre* transformations, as shown below.

We shall write the *Gibbs* fundamental equation in the conventional form

$$T\,\mathrm{d}S = \mathrm{d}U + p\,\mathrm{d}V - \mu\,\mathrm{d}N \tag{4.5}$$

instead of eq. (4.1). Here we dropped the summation dummies l and a as indices at p_l, V^l and μ_a, N^a. The extensity V stands for all working parameters V^l, and N for all particle numbers N^a. We now can obtain different thermodynamic potentials as listed below. The corresponding natural variables are the arguments written in the brackets. The arguments which have to be kept constant by performing the derivatives are marked as suffixes if this is advantageous.

Entropy $S(U, V, N)$ yields:

$$\left(\frac{\partial S}{\partial U}\right)_{VN} = \frac{1}{T}, \tag{4.6}$$

$$\left(\frac{\partial S}{\partial V}\right)_{UN} = \frac{p}{T}, \tag{4.7}$$

$$\left(\frac{\partial S}{\partial N}\right)_{UV} = -\frac{\mu}{T}. \tag{4.8}$$

Internal energy $U(S, V, N)$:

$$\left(\frac{\partial U}{\partial S}\right)_{VN} = T, \tag{4.9}$$

$$\left(\frac{\partial U}{\partial V}\right)_{SN} = -p, \tag{4.10}$$

$$\left(\frac{\partial U}{\partial N}\right)_{SV} = \mu. \tag{4.11}$$

Helmholtz free energy $F(T, V, N)$: The definition

$$F = U - TS \tag{4.12}$$

can be read as *Legendre* transformation from the independent variable S to T. This yields directly:

$$\left(\frac{\partial F}{\partial T}\right)_{VN} = -S, \tag{4.13}$$

$$\left(\frac{\partial F}{\partial V}\right)_{TN} = -p, \tag{4.14}$$

$$\left(\frac{\partial F}{\partial N}\right)_{TV} = \mu. \tag{4.15}$$

It should be stressed that for example the derivatives in eq. (4.10) and (4.14) are two different operations because the variables kept constant are different.

Enthalpy $J(S, p, N)$: It is defined by

$$J = U + pV, \tag{4.16}$$

which is a *Legendre* transformation from the variable V to p:

$$\left(\frac{\partial J}{\partial S}\right)_{pN} = T, \tag{4.17}$$

$$\left(\frac{\partial J}{\partial p}\right)_{SN} = V, \tag{4.18}$$

$$\left(\frac{\partial J}{\partial N}\right)_{Sp} = \mu. \tag{4.19}$$

The usefulness of the introduction of the enthalpy can be seen if we remind that the heat added reversibly to a system is

$$đQ = T\,dS. \tag{4.20}$$

In a process with constant V and N,

$$đQ = dU. \tag{4.21}$$

In a process with constant p and N, however,

$$đQ = dJ. \tag{4.22}$$

In practical cases, constant intensities p are frequent constraints. For instance, chemical reactions are more frequently observed at constant atmospheric pressure than in closed vessels. This leads to the following thermodynamic potential.

Gibbs free energy $G\,(T, p, N)$: The definition is

$$G = F + pV = J - TS. \tag{4.23}$$

As a *Legendre* transformation this yields

$$\left(\frac{\partial G}{\partial T}\right)_{pN} = -S, \tag{4.24}$$

$$\left(\frac{\partial G}{\partial p}\right)_{TN} = V, \tag{4.25}$$

$$\left(\frac{\partial G}{\partial N}\right)_{Tp} = \mu. \tag{4.26}$$

The advantage of the use of G is that the natural variables T, p are intensities and therefore controlled by the environment.

All intensities as natural variables. The choice of all intensities as independent variables leads to the *Legendre* transformation

$$\Phi = G - \mu N. \tag{4.27}$$

The corresponding thermodynamic potential $\Phi\,(T, p, \mu)$ has no special name. Introducing Φ yields

$$\left(\frac{\partial \Phi}{\partial T}\right)_{p\mu} = -S, \tag{4.28}$$

$$\left(\frac{\partial \Phi}{\partial p}\right)_{T\mu} = V, \tag{4.29}$$

$$\left(\frac{\partial \Phi}{\partial \mu}\right)_{Tp} = -N. \tag{4.30}$$

In chapter 2.3.7 we shall see that the thermodynamic potentials have yet another distinguishing property: They become extreme in equilibrium if the natural variables are held fixed by external contraints. This is another analogy to mechanical potentials. Examples of the use of the thermodynamic potentials for practical purposes will be given in chapter 2.3.7.

The *Gibbs-Duhem* equation. A system which has the same properties in all its spatial parts is called "homogeneous". This includes not only that the material composition and the dynamical laws anywhere in the interior are the same, but also that the thermal state of all parts, described by the intensities, is the same. For instance a "fluid medium", that is a gas or a liquid, can be in such a state if no surface tension and no external forces like gravitation have to be taken into account. When we say that all working parameters V^l of the system are genuine extensities, we mean the following. Let us compare a spatial part αV of volume V with the whole system, then αV^l have to be the working parameters of this part when V^l are the corresponding values of the whole homogeneous system in the same thermal state. In phenomenological theory, internal energy U and entropy S are understood to be such genuine extensities as well. In a homogeneous system thus for any $0 < \alpha \leqslant 1$ it holds that

$$\alpha S\,(U,\,V,\,N) = S\,(\alpha U,\,\alpha V,\,\alpha N). \tag{4.31}$$

In this context we prefer the explicit use of the summation indices l and a. By differentiation with respect to α and putting $\alpha = 1$ afterwards, we obtain

$$S = \frac{\partial S}{\partial U}\,U + \frac{\partial S}{\partial V^l}\,V^l + \frac{\partial S}{\partial N^a}\,N^a. \tag{4.32}$$

Due to eqs. (4.6, 4.7, 4.8) this means

$$TS = U + p_l\,V^l - \mu_a\,N^a, \tag{4.33}$$

or with eqs. (4.12, 4.33):

$$G = \mu_a\,N^a. \tag{4.34}$$

This relation is called the *Gibbs-Duhem equation*.

In the following we shall designate an extensity, like G, by the corresponding small letter (like g) instead of the capital if taken for one mole. For a pure substance, eq. (4.34) then yields:

$$g = \mu. \tag{4.35}$$

If a system composed of homogeneous subsystems only, which we designate by the subscript i, is in absolute equilibrium, any extensity will be the sum of the values of this extensity in the subsystems. The intensities, however, will be equal in all subsystems. This holds in particular for the μ_a. Thus we obtain

$$G = \mu_a \sum_i N_i^a = \mu_a\,N^a. \tag{4.36}$$

This means:

> The *Gibbs-Duhem* equation is valid also for a system composed of homogeneous subsystems when in absolute equilibrium.

2.3.5 Thermal and Caloric Equations of State

In the preceding chapter we demonstrated the distinguished role of the "natural" variables associated with the respective thermodynamic potential. In the following we shall show that we can obtain other efficacious relations by the use of other independent variables than the natural ones. The root again will be the *Gibbs* fundamental equation.

As already pointed out, we can describe a thermal state by the set M of all extensities M^ν or by the set λ of all intensities λ_ν. Of course we can take also a mixed set if the number of the independent variables is the same. The equations which connect the independent thermal variables with one another are called *equations of state*. As internal energy U, and with this the thermally conjugate temperature T, are of particular significance,

$$U = U(T, y) \tag{5.1}$$

is conventionally called the *caloric equation of state*, whereas relations of the form

$$M^l = M^l(T, y) \tag{5.2}$$

are called *thermal equations of state*. These names are used likewise if the independent variables are chosen not to be y but M or mixed sets.

A standard example is the *ideal gas*. This is a gas sufficiently rarefied at not extremely low temperature. The thermal equation of state then is uniform for different chemical species if the amount is one mole:

$$v = \frac{RT}{p}, \tag{5.3}$$

where v is volume of one mole, p is pressure, and R is a universal constant, the so-called *gas constant*

$$R = 1.98 \text{ cal/grade mole.} \tag{5.4}$$

Eq. (5.3) is called the *ideal gas equation*.

We now return to the general case. In the *Gibbs* fundamental equation

$$T\,dS = dU + y_l\,dM^l \tag{5.5}$$

we can choose T, M as independent variables which are "natural" variables neither of S nor of U. We then obtain:

$$\left(\frac{\partial S}{\partial T}\right)_M = \frac{1}{T}\left(\frac{\partial U}{\partial T}\right)_M, \tag{5.6}$$

$$\left(\frac{\partial S}{\partial M^l}\right)_T = \frac{1}{T}\left(\frac{\partial U}{\partial M^l}\right)_T + \frac{y_l}{T}. \tag{5.7}$$

As $S(T, M)$ and $U(T, M)$ are unique functions, their cross derivatives of the second order with respect to T, M are independent of the sequence of the differentiations. We obtain

$$\left(\frac{\partial U}{\partial M^l}\right)_T = -y_l + T\left(\frac{\partial y_l}{\partial T}\right)_M. \tag{5.8}$$

This is a connection between the thermal and the caloric equations of state and a remarkable consequence of the Second Law.

If we choose T, y as independent variables, it is convenient to change from U to the *Legendre* transform

$$\hat{J} = U + y_l M^l. \tag{5.9}$$

We can call this quantity a generalized enthalpy. With this the *Gibbs* fundamental equation takes the form

$$T\,\mathrm{d}S = \mathrm{d}\hat{J} + M^l\,\mathrm{d}y_l \tag{5.10}$$

instead of eq. (5.8) we obtain the relation

$$\left(\frac{\partial \hat{J}}{\partial y_l}\right)_T = M^l - T\left(\frac{\partial M^l}{\partial T}\right)_y. \tag{5.11}$$

The result of eqs. (5.7, 5.8) is

$$\left(\frac{\partial S}{\partial M^l}\right)_T = \left(\frac{\partial y_l}{\partial T}\right)_M. \tag{5.12}$$

By the use of $\hat{J}(T, y)$ instead of $U(T, M)$ we obtain in an analogous way

$$\left(\frac{\partial S}{\partial y_l}\right)_T = -\left(\frac{\partial M^l}{\partial T}\right)_y. \tag{5.13}$$

These equations are called *Maxwell relations*.

The Gas. To give an illustration, we consider a gas, not necessarily "ideal". The only M^l is volume V with the thermal conjugate pressure p. Eq. (5.8) reads

$$\left(\frac{\partial U}{\partial V}\right)_T = -p + T\left(\frac{\partial p}{\partial T}\right)_V. \tag{5.14}$$

For the ideal gas this expression is zero, which means that U is independent of volume and a function only of temperature T. In the history of thermodynamics this fact is of particular interest. The ideal gas equation (5.3) was known earlier than the Second Law. Therefore the connection eq. (5.14) between thermal and caloric equation of state was unknown as well. It was found empirically, however, that U of the ideal gas is independent of volume. Therefore, in the theoretical scheme, the ideal gas was characterized by two properties, then believed to be independent: the ideal gas equation and the volume independence of U. With the Second Law it became obvious, that the thermal equation of state is sufficient for determining the thermodynamic properties of the ideal gas.

The Radiation Cavity. Instead of the gas we can consider a cavity filled with electromagnetic radiation. There are equilibrium states with the environment which has a certain temperature T. In such a state the average absorption and emission of radiation at the walls of the cavity are equilibrated with each other. The temperature of the cavity is defined as equal to that of the walls. Indeed, any piece of matter brought into the cavity will also accept this temperature in the thermal equilibrium. The spatial density of the radiation energy, which we call u, is homogeneous in the cavity. The radiation acts on the walls by a so-called *radiation pressure p*. The total energy of the cavity is

$$U = uV. \tag{5.15}$$

Eq. (5.14) therefore yields

$$\frac{dp}{dT} = \frac{u+p}{T}\,.$$

(5.16)

Due to electromagnetic theory the relation

$$p = \frac{u}{3}$$

(5.17)

holds. We shall obtain it in chapter 4.1.6. Wit eq. (5.16) it leads to the so-called *Stefan-Boltzmann law*

$$u = CT^4\,.$$

(5.18)

We shall demonstrate in chapter 3.3.2 that the so-called *Stefan-Boltzman constant C* is universal, i.e. independent of the material of the walls. This will also turn out to be a consequence of the Second Law.

2.3.6   Availability

In this chapter we shall discuss to concept of *availability* also called *exergy*. The importance of this quantity was already emphasized by *Gibbs*[10]. Nevertheless it found adequate notice in applied thermodynamics to a larger extent not earlier than in the last decades[11]. In this book it will play a prominent role in several considerations.

Maximum available work. In the following we consider a system not in equilibrium with its environment. The environment shall be in an equilibrium state as well, and so large that its intensities remain unchanged during the state changes of the system. The system called Σ can exchange work and heat with its environment Σ^* and, moreover, it can give pure work energy $\tilde{W}$ to a third external system (see Figure 12). The changes ΔM^{*l}, ΔU^* (final minus initial values) of the large environment may be assumed to be

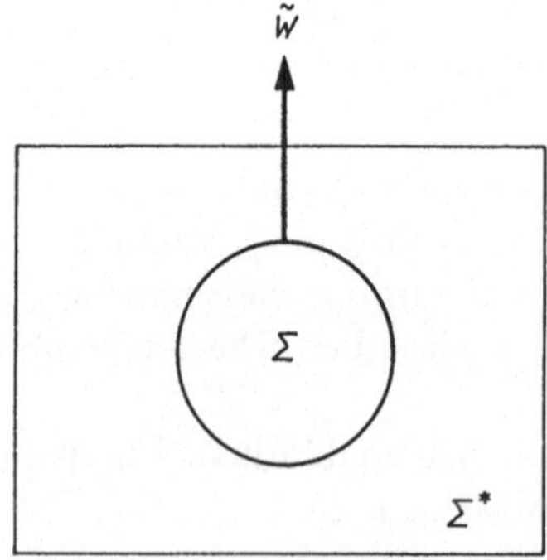

Figure 12

Illustration of the definition of "availability".
External work $\tilde{W}$ available from a system Σ in a heat bath Σ^*.

[10] *J. W. Gibbs*, Collected Works, Vol. 1 (Yale University, New Haven 1948).

[11] *J. H. Keenan*, Thermodynamics (MIT Press, Cambridge, Mass. 1970).

quasistatic finite changes of conserved quantities M^l and energy U. With the changes $\Delta M^l, \Delta U$ of the system Σ then

$$\Delta M^l + \Delta M^{*l} = 0, \tag{6.1}$$

$$\Delta U + \Delta U^* = -\widetilde{W}. \tag{6.2}$$

The compound system formed by Σ plus Σ^* is adiabatically closed. Therefore the Second Law yields

$$\Delta S + \Delta S^* \geqslant 0. \tag{6.3}$$

The intensities of Σ^* shall be T^0, y^0. As they remain unchanged, with

$$\Delta S^* = \frac{1}{T^0} (\Delta U^* + y_l^0 \Delta M^{*l}), \tag{6.4}$$

for the gained work we obtain the inequality

$$\widetilde{W} \leqslant -\Delta U + T^0 \Delta S - y_l^0 \Delta M^l = -\Delta \Lambda. \tag{6.5}$$

The maximum available work is thus the decrease of the so-called *availability* or *exergy*

$$\Lambda = U - U^0 + y_l^0 (M^l - M^{0l}) - T^0 (S - S^0). \tag{6.6}$$

The superscript zero designates the values belonging to the equilibrium state of Σ with Σ^*. The availability is defined to vanish in this state. According to the Second Law, in absolute equilibrium no work $\widetilde{W}$ can be gained. Otherwise the system would be a perpetuo mobile of second kind. To bring Σ out of the equilibrium is possible only if work is done on Σ. This means that Λ is never negative. We write

$$\Lambda = U - U^0 + p^0 (V - V^0) - \mu^0 (N - N^0) - T^0 (S - S^0) \geqslant 0 \tag{6.7}$$

if all working parameters V^l are conserved quantities (like volume or electrical charge) and if particle numbers N^a vary only through exchange with the environment, not because of chemical reactions.

Entropy production. Another result of the Second Law can be obtained by application of the inequality (6.5) to the case that there is no contact with a third system, i.e. that $\widetilde{W}$ vanishes. Then we call the changes of the system in time *spontaneous* changes and obtain:

> Availability never increases spontaneously.

For these spontaneous changes, in eq. (6.5), we can separate ΔS into two terms of different character. The first one,

$$\Delta S_{ex} = \frac{1}{T^0} (\Delta U + y_l^0 \Delta M^l), \tag{6.8}$$

is entropy transferred from the environment Σ^* by mere exchange. It is an *entropy flow* and vanishes if no exchange of U and M between Σ and Σ^* is present. — The second term,

$$\Delta S_{pr} = -\frac{1}{T^0} \Delta \Lambda \geqslant 0, \tag{6.9}$$

is entropy produced in the interior of the system Σ, sometimes also called *irreversibility*. The expression of eq. (6.9) is entropy produced by the change ΔU, ΔM. Entropy produced by the whole equilibration of the system to the mutual equilibrium with the environment is Λ/T^0.

By convention, the entropy produced in the interior of the system given by eq. (6.9) is called *entropy production* Θ if related to the unit of time. The inequality in (6.9) thus yields:

$$\Theta = -\frac{1}{T^0}\,\dot{\Lambda} \geqslant 0. \tag{6.10}$$

This is the important result:

> Entropy production is never negative if the system is in an environment with unchanged intensities.

Availability and information gain. Whereas we have in this section argued up to now strictly in the framework of phenomenological thermodynamics, we shall now make an intermission and shall return to the statistical theory for the rest of this chapter. We compare availability Λ with information gain

$$K(\rho, \rho^0) = \mathrm{tr}\,[\rho\,(\log \rho - \log \rho^0)], \tag{6.11}$$

which can also be written in the form

$$K(\rho, \rho^0) = I(\rho) - I(\rho^0) - \mathrm{tr}\,[(\rho - \rho^0)\log \rho^0]. \tag{6.12}$$

Let

$$\rho^0 = \exp\frac{1}{kT^0}\,(\Phi^0 - \mathbf{H} - y^0\mathbf{M}) \tag{6.13}$$

be the state of the system Σ if in the mutual equilibrium with the environment Σ^*, and let ρ be the momentary state of Σ, then we obtain

$$K(\rho, \rho^0) = -(S - S^0) + \frac{1}{T^0}\,(U - U^0) + \frac{y^0}{T^0}\,(M - M^0). \tag{6.14}$$

This means, that

$$K(\rho, \rho^0) = \frac{\Lambda}{kT^0}\,. \tag{6.15}$$

Availability (exergy) is, up to an unimportant multiplier, information gain, which appears already in general statistics as a fundamental quantity, not only in physics. Positivity of availability is identical with the fundamental principle of positivity of information gain.

It should be mentioned that in statistical thermodynamics we do not need the restriction that the occurring $\mathbf{M}$ are strictly conserved quantities. Moreover, ρ can be any probability distribution in Γ-space, not only a generalized canonical one of the same type like ρ^0. So we see that the statistical foundation of inequality (6.6) is more general than the phenomenological one.

The principle that "produced entropy" never can decrease in time if the system is in an environment with unchanged intensities, in the statistical interpretation assumes the form

$$\frac{\mathrm{d}}{\mathrm{d}t} K(\rho, \rho^0) \leqslant 0. \tag{6.16}$$

We have to analyse this in the statistical description. ρ depends on time t, yet ρ^0 does not. The knowledge of the dates of the environment is comprised in ρ^0 and remains unchanged. The excess knowledge about the microstates of the system is measured by the information gain $K(\rho, \rho^0)$. Inequality (6.16) reads:

> The excess knowledge cannot increase after the last observation.

If information gain is accepted to be the adequate measure of the excess knowledge, then this seems to be a very general postulate which we have to require for any adequate dynamics. The Second Law then appears as this principle in the special case that ρ^0 is the homogeneous distribution in phase space.

We can obtain the principle also without appealing to the given interpretation of information gain by using the relation eq. (3.12) of chapter 1.2.3. The macrostate ρ shall be given with the probability distribution P over the microstates i. Let us represent this distribution by a large ensemble of independent systems in different microstates i. Let moreover Λ_{ji} be the transition probability of each system of going in a certain short time from the microstate i to the microstate j. Then eq. (3.6) of chapter 1.2.3 gives the change of the probability distribution P to $\hat{P}$ in this time. As the members of the ensemble are independent from one another, the transition probabilities Λ_{ji} are not dependent on the distribution P and relation eq. (3.12) is fulfilled. This means that $K(P, P^0)$ never increases in time, according to eq. (6.16).

As explained in chapter 2.2.4, it is possible to use different distributions ρ^0 for the same macroscopic thermal state. Correspondingly, we can obtain different types of availability. For instance we may use the canonical distribution in which volume is a sharp parameter and obtain

$$\Lambda = U - U^0 - T^0(S - S^0) \geqslant 0. \tag{6.17}$$

Or we may use the pressure ensemble in which volume V is a fluctuating random quantity and obtain

$$\Lambda = U - U^0 - T^0(S - S^0) + p^0(V - V^0) \geqslant 0. \tag{6.18}$$

This may be continued by proceeding to further distributions as described in the above-mentioned chapter 2.2.4. The same manifold of inequalities can be obtained in the phenomenological framework by changing the set M in eq. (6.7).

Reaction heat and chemical activity. To give an illustration of the availability, we shall now discuss an important application of the developed concepts and speak of chemical reactions which occur in a mixture of different chemical components. They are supposed to occur in a closed vessel so that the system Σ cannot exchange matter with its environment Σ^*. Then

$$\Delta \Lambda = \Delta U + p^0 \Delta V - T^0 \Delta S. \tag{6.19}$$

First we assume that the volume V of the vessel is fixed, and no work is done by the reactor. This is the case if the reaction takes place in a so-called *Berthelot bomb*. The reaction then is called *isochoric*. The reaction has to be complete. This means that we compare an initial state 1 in which only the initial reactants are present, with a final state 2 in which only the final products of the reaction are present. Both shall be equilibria with the temperature T^0 of the environment. To realize this situation, we have to wait a sufficient long time for the system to be able to assume again a thermal equilibrium with the environment after the reaction. The reaction then is called *isothermal*. For the maximum available work we obtain

$$\Delta \Lambda = F_2 - F_1 = \Delta F, \tag{6.20}$$

where F is *Helmholtz* free energy. In most cases this amount of energy is observed in the form of heat and not of work. Then the total energy

$$-\Delta U = -\Delta F + T^0 \Delta S = Q_V \tag{6.21}$$

produced by the reaction appears as heat given to the environment. Therefore it is called isochoric *reaction heat* if related to a standard amount of moles. It should be stressed, however, that this name is misleading because the part ΔF can indeed be obtained as work. This is possible for instance if the reaction occurs in a galvanic element. By a counter-voltage the velocity of the reaction can be controlled and made arbitrarily slow, so that the reaction can approximate a quasistatic process. The available work can be gained as work of the electrical current. The remaining part $T \Delta S$ of $-\Delta U$ is called *bound energy* because it is not available as work. This name is used for the product TS too.

If the reaction heat Q_V, which is the total energy produced by the reaction, is positive, the reaction is called *exothermal*. It is called *endothermal* if the reaction heat is negative. Then the reaction needs energy withdrawn form the environment in the form of heat.

In most cases, chemical reactions are not observed at fixed volume but rather at fixed pressure. Then they are called *isobaric*. For a theoretical discussion it is useful to visualize the reaction occurring in a vessel with movable walls, say in form of a piston. Again, after a certain time the system will come into a final thermal equilibrium with the environment and assume the same pressure and temperature as in the initial state. Then ΔT and Δp are zero and with *Gibbs* free energy G we obtain for the maximum available work

$$\Delta \Lambda = G_2 - G_1 = \Delta G. \tag{6.22}$$

Then not $-\Delta U$ but the decrease $-\Delta J$ of enthalpy

$$J = U + pV \tag{6.23}$$

is called *reaction heat*, or explicitly the *isobaric reaction heat* Q_p if again related to the standard amount of moles. The system has to perform the work $p \, \Delta V$ against the environment by shifting the piston. Q_p is the sum of the decrease Q_V of the internal energy and this work. This reaction heat can be either positive or negative. In the first case the isobaric reaction is called exothermal, in the second case endothermal. It is quite possible that the same reaction is exothermal for constant volume and endothermal for constant pressure and vice versa. In our discussion we have tacitly assumed that the reactor could only perform work by change of the volume. To generalize the result to more working parameters, we have to replace products pV with the sum $p_l V^l$, as was explained in chapter 2.2.2. For instance, in case of electrical work in a galvanic element, V^l is charge, and $-p_l$ is voltage.

In an early state of thermodynamics it was believed that reaction heat was a measure for the tendency of a reaction. This, however is not the case. Also endothermal reactions can pass spontaneously. Since the importance of the Second Law was realized, the availability is considered the adequate measure for the tendency of a reaction. Availability of a chemical reaction for the standard amount of moles therefore is called the *activity* of the reaction. Nevertheless it should be noticed that the activity is not a measure of how fast a reaction passes off. The reaction may pass in the form of a vehement detonation in the presence of a catalyst or an ignition; but without such a presence it might not happen for years. The activity, however, is the same.

We have to distinguish between activity A_V of isochoric and activity A_p of isobaric reactions. The first one is the decrease $-\Delta F$ of *Helmholtz* free energy, the second one $-\Delta G$ of *Gibbs* free energy caused by the reaction of the standard amount of moles. We shall return to these concepts in more detail in section 3.2.

Eqs. (6.20, 6.22) are typical for a more general connection which will be observed in the following chapters. Availability and information gain are dependent on two states. External constraints on the system like the fixing of volume or pressure can result in making these quantities equal the difference of a function of only one state in these two states. This feature will also distinguish these functions as thermodynamic potentials.

2.3.7 Equilibrium Conditions

In this and in the next chapter we shall consider relations which can be deduced directly from the positivity of availability. These relations will show that thermal equilibria are distinguished by the extremum of a thermodynamic potential. Of which one, this depends on the restrictions the system is subject to. The corresponding relations are of two different types. Of the first type are equations corresponding to vanishing derivatives in an extremum. It will turn out that they are collected in the *Gibbs* fundamental equation. The relations of the second type are inequalities determining the sign of second derivatives in an extremum. These inequalities express the stability of the equilibrium and will be discussed in the next chapter.

The following discussion will be simpler if we return to the designations λ_ν, M^ν that we originally used for the thermal variables. Therefore we write the availability again in the form

$$\Lambda = kT^0 K(\rho, \rho^0),\tag{7.1}$$

yet without needing any reference to the statistical concept of information gain. In this context let K be nothing else besides the macroscopic expression

$$K(\rho, \rho^0) = I - I^0 + \lambda_\nu^0 (M^\nu - M^{0\nu}),\tag{7.2}$$

and ρ, ρ^0 be mere symbols for two thermal states. With eq. (4.1)

$$\frac{\partial I}{\partial M^\nu} = -\lambda_\nu \tag{7.3}$$

we obtain

$$\frac{\partial}{\partial M^\nu} K(\rho, \rho^0) = -(\lambda_\nu - \lambda_\nu^0). \tag{7.4}$$

We know that any increase of availability Λ and thus of K requires a positive work $\widetilde{W}$ done on the system. Therefore the right hand side of eq. (7.4) represents a *resistance* against a deviation of M^ν from the equilibrium value. This resistance, which we shall also call a restoring *force* in a metaphorical sense, exemplifies the tendency of the system to go into the equilibrium ρ^0 with the environment by "spontaneous" changes. Eq. (7.4) has some similarity with the connection of this force with a potential K in mechanics.

The comparison with a potential, however, fails in so far as K depends on two states ρ, ρ^0 and not on the the momentary state ρ of the system alone. Nevertheless the analogy of a potential can be obtained if the system is subject to restrictive conditions. We shall see moreover that anyone of the already introduced thermodynamic potentials distinguishes the equilibrium ρ^0 by becoming an extremum.

To this purpose we now start with the positivity of availability written in the form

$$\Lambda = (U - U^0) + p^0(V - V^0) - \mu_a^0(N^a - N^{0a}) - T^0(S - S^0) \geqslant 0. \tag{7.5}$$

By going over to the limit that the differences in the brackets become infinitesimal we obtain an equality. It is the *Gibbs* fundamental equation. We also consider, however, finite values of the brackets. Let volume V be the only working parameter. A generalization to more working parameters is easily obtained.

In an energetically and materially isolated system with constant volume all extensities U, V, N^a remain unchanged. Thus

$$S - S^0 \leqslant 0. \tag{7.6}$$

This means that entropy is maximum in equilibrium. Availability hereby becomes again the difference of two values of a one-state function by the restrictions.

We can write eq. (7.5) also in other forms as

$$\Lambda = (F - F^0) + p^0(V - V^0) - \mu_a^0(N^a - N^{0a}) + S(T - T^0) \geqslant 0, \tag{7.7}$$

$$\Lambda = (G - G^0) - V(p - p^0) - \mu_a^0(N^a - N^{0a}) + S(T - T^0) \geqslant 0. \tag{7.8}$$

If the system is subject to the restrictions that temperature and volume are constant, inequality (7.7) yields

$$F - F^0 \geqslant 0. \tag{7.9}$$

This means that the *Helmholtz* free energy F is minimum in equilibrium with the environment.

Similarly we can obtain different results corresponding to different restrictive conditions. To sum them up, we can say, equilibrium is distinguish by

$$S \;= \text{max for constant } U, V, N,$$
$$U = \text{min for constant } S, V, N,$$
$$J \;= \text{min for constant } S, p, N,$$
$$F \;= \text{min for constant } T, V, N,$$
$$G = \text{min for constant } T, p, N,$$
$$\Phi = \text{min for constant } T, p, \mu.$$

This cedes the following rule:

> The equilibrium is distinguished by the extremum of a thermodynamic potential if the corresponding natural variables are hold constant by restrictive conditions.

We already recognized an analogy to mechanical potentials in chapter 2.3.4 which justified the name thermodynamic "potential". Now we found another analogy, the extremum property in equilibrium.

Vapor pressure. To give an illustration how this extremum property of a thermodynamic potential works in practical cases, we shall discuss a special example. Under certain conditions the liquid and the vapor of the same substance, say water, are in thermal equilibrium with each other. The corresponding equilibrium pressure is called *vapor pressure P*. It is dependent on temperature T only. Any change of volume of the system, which consists partially of liquid and partially of vapor, will lead to a transition of one phase into the other until vapor pressure $P(T)$, and equilibrium is reached again.

Let the molar *Gibbs* energy of liquid be called g', and of vapor g'', respectively. N', N'' shall be the corresponding mole numbers. The total *Gibbs* free energy is

$$G = N'g' + N''g''. \tag{7.10}$$

By the transition of $\Delta N'$ mole liquid into vapor

$$\Delta N' + \Delta N'' = 0. \tag{7.11}$$

If the change occurs by vaporization at constant vapor pressure, we obtain

$$\Delta G = (g' - g'')\Delta N' = 0 \tag{7.12}$$

because *Gibbs* free energy G is minimum in equilibrium. This yields

$$g'(T, P(T)) = g''(T, P(T)). \tag{7.13}$$

Differentiation with respect to T and the use of eqs. (4.24, 4.25)

$$\left(\frac{\partial g}{\partial T}\right)_p = -s, \qquad \left(\frac{\partial g}{\partial p}\right)_T = v \tag{7.14}$$

gives

$$\frac{dP}{dT} = \frac{s'' - s'}{v'' - v'}. \tag{7.15}$$

This is called the *Clausius Clapeyron equation*:

$$\frac{dP}{dT} = \frac{q}{(v'' - v')T} \tag{7.16}$$

where

$$q = (s'' - s')T \tag{7.17}$$

is the *heat of vaporization* of one mole. Umlike the chemical reaction heat, q is defined positive by convention if it has to be given to the system when the "reaction", here the evaporation, takes place. It might be mentioned that eq. (7.15) is valid for any phase transition of a homogeneous substance with a finite change of volume, for instance for the melting of a solid.

Eq. (7.16) can be integrated easily in a familiar approximation applicable if T is not too close to the critical temperature at which the dissimilarity between liquid and vapor vanishes. In this approximation the volume v' of the liquid is neglected in comparison with the volume v'' of the vapor. Moreover, the vapor is assumed to satisfy the ideal gas equation

$$v'' = \frac{RT}{P(T)}. \tag{7.18}$$

q is assumed not to depend considerably on T in the regime of application. Then the integration yields:

$$P(T) = C \exp\left(-\frac{q}{RT}\right). \tag{7.19}$$

Vapor pressure of droplets. Hitherto we were considering a plane surface of the liquid. In a droplet, however the surface tension σ gives rise to an energy contribution and changes the vapor pressure. The surface ω is a working parameter of exceptional character because it is not proportional to volume, thus not a "genuine" extensity in the restricted sense. An increase $d\omega$ of the surface requires the work

$$đW = \sigma \, d\omega \tag{7.20}$$

if performed in a quasistatic way. Let us consider a spherical droplet with radius r. A change of volume V' and a change of surface ω are connected with each other by

$$r \, d\omega = 2 \, dV'. \tag{7.21}$$

To find the influence of the surface on the pressure in the droplet, we study the case that droplet and vapor are enclosed in a vessel of fixed volume. Now T, V, N are kept constant and therefore the *Helmholtz* free energy F is minimum in the equilibrium:

$$dF = -p' dV' - p'' dV'' + \sigma \, d\omega = 0. \tag{7.22}$$

With

$$dV = dV' + dV'' = 0 \tag{7.23}$$

this yields

$$dF = \left(p'' - p' + \frac{2\sigma}{r}\right)dV'' = 0. \tag{7.24}$$

In equilibrium thus

$$p' = p'' + \frac{2\sigma}{r}. \tag{7.25}$$

ω and V' of the spherical droplet are not independent working parameters. Therefore the intensity p is not the same in the droplet and on the outside. This is an example in which the conditions for the theorem of equal intensities, as quoted in chapter 2.1.6, are not fulfilled: p and σ are not independent intensities.

The pressure p' in the droplet is higher than the pressure p'' outside, which is vapor pressure $P(T)$. The smaller the droplet, the larger the difference of these values. Let us now consider the vaporization of dN' moles from the spherical surface at constant vapor pressure. Not volume but pressure is kept constant. Equilibrium is distinguished again by

$$dG = (g' - g'')\,dN' = 0, \tag{7.26}$$
$$g'(T, p') = g''(T, p''). \tag{7.27}$$

Differentiation of the last equation with respect to r at fixed temperature results in

$$\left(\frac{\partial g'}{\partial p'}\right)_T \left[\left(\frac{\partial P}{\partial r}\right)_T - \frac{2\sigma}{r^2}\right] = \left(\frac{\partial g''}{\partial p''}\right)_T \left(\frac{\partial P}{\partial r}\right)_T. \tag{7.28}$$

With eq. (7.13) thus

$$\left(\frac{\partial P}{\partial r}\right)_T = -\frac{2\sigma}{r^2}\frac{v'}{v'' - v'}. \tag{7.29}$$

In the same approximation as explained before,

$$v'' - v' \approx v'' = \frac{RT}{P}, \tag{7.30}$$

we obtain

$$\ln\frac{P}{P_\infty} = \frac{2\sigma v'}{RTr}, \tag{7.31}$$

where P_∞ is the vapor pressure over the plane surface $r = \infty$. The result is plotted qualitatively in Figure 13. We now compare droplets of different size in the vapor of a given pressure p_0. With this external pressure a *critical radius* r_0 of those droplets is associated for which P is equal to p_0. Smaller droplets will vaporize and vanish. Droplets with r larger than r_0 will condense and increase. If they are spherical, only these droplets form *nuclei of condensation*.

Gibbs phase rule. Now we contemplate a system compound of Γ chemical *components* in Φ *phases*. The "phases" are homogeneous parts of the system which are spatially separated from one another. All such parts which are of the same consistence

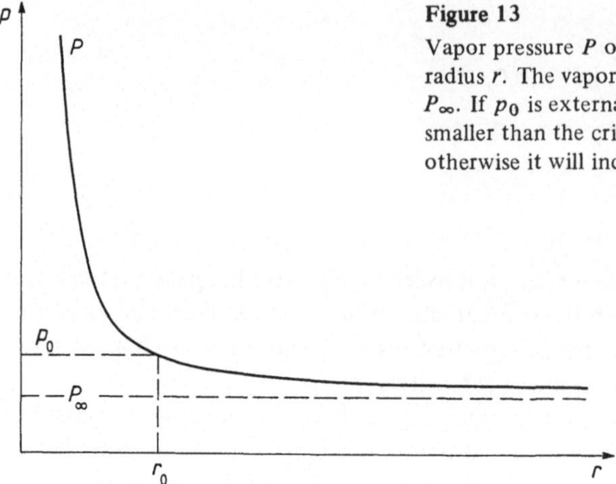

Figure 13

Vapor pressure P of a droplet as a function of the radius r. The vapor pressure over a plain surface is P_∞. If p_0 is external pressure, a droplet with r smaller than the critical radius r_0 will evaporate; otherwise it will increase with condensation.

form "one phase", like for instance crystal pieces of the same material composition. To give an example, the components may be water and ammonium chloride. Possible phases are: a gaseous mixture, a liquid mixture, and three solid phases, namely ice crystals, chloride crystals, and crystallized mixture. In other systems also more liquid phases than one are possible, like oil in water. We assume in the following that the spatial extension of any part is so large that surface tension can be neglected.

The question we shall be concerned with is how many thermal variables are free in a thermal equilibrium of such a system. The intensities are T, p and all chemical potentials μ_b^a of the components a in the phase b. Subscript a runs from 1 to Γ, superscript b runs from 1 to Φ. The chemical components may be mixed with one another. It is, however, supposed that no chemical reactions occur between the components and thus that their mole numbers N^a remain unchanged. In the thermal equilibrium, *Gibbs* free energy is minimum. This yields

$$dG = \sum_{ab} \mu_b^a \, dN_b^a = 0 \tag{7.32}$$

on the restriction that

$$dN^a = \int_b dN_b^a = 0 \tag{7.33}$$

for any component a. With *Lagrange* multipliers τ^a thus:

$$\sum_{ab} (\mu_b^a - \tau^a) \, dN_b^a = 0. \tag{7.34}$$

This gives for any component a

$$\mu_b^a = \tau^a \tag{7.35}$$

and can be written as

$$\mu_1^a = \mu_2^a = \ldots = \mu_\Phi^a. \tag{7.36}$$

These are $\Phi - 1$ equations for each component; for all Γ components altogether there are $\Gamma(\Phi - 1)$ equations. The thermal variables we can easily control, are T, p and the independent relative concentrations of the components. In one phase the number of all relative concentrations is $\Gamma - 1$ and therefore $\Phi(\Gamma - 1)$ in all Φ phases. The thermal variables are not independent from one another but restricted by $\Gamma(\Phi - 1)$ equations. The independent variables are called *degrees of freedom*. Their number is

$$f = 2 + \Phi(\Gamma - 1) - \Gamma(\Phi - 1). \tag{7.37}$$

This is *Gibbs phase rule*:

$$f = \Gamma - \Phi + 2. \tag{7.38}$$

Expressed in words:

> The number of degrees of freedom in the equilibrium of a compound system is the number of components minus the number of phases plus two.

Let us give some examples:

(a) For a *gas of pure substance* $\Gamma = \Phi = 1; f = 2$. Indeed the two variables can be chosen arbitrarily in a certain regime and make an equilibrium possible.

(b) In the coexistence of *gas and liquid* of the same substance, $\Gamma = 1$, $\Phi = 2$; $f = 1$. Only T is free and determines vapor pressure $P(T)$. Likewise we can say that pressure is free but boiling temperature then is determined.

(c) In the system of *gas, liquid, and solid* of the same substance, $\Gamma = 1$, $\Phi = 3$; $f = 0$. In Figure 14 the diagram is drawn schematically. The three domains of gas, liquid, and solid state are designated by the letters g, l, and s respectively. They are separated by coexistence curves. The *sublimation* curve separates s from g. It starts in the zero

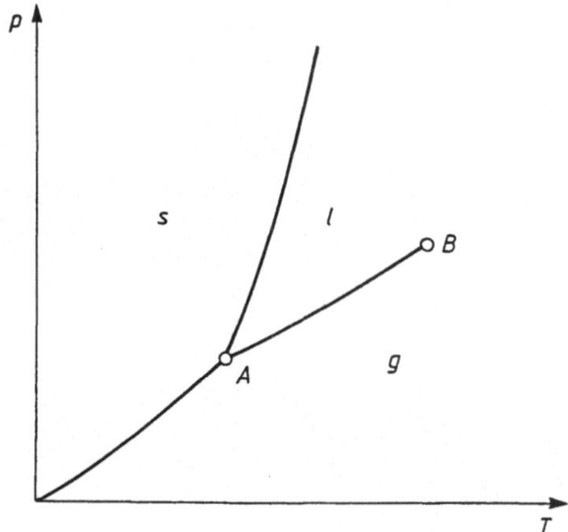

Figure 14

State diagram of the gas (g), liquid (l), and solid (s) in the T, p-plane. A is the triple point, and B the critical point.

point and ends in the *triple point* A. The *melting* curve separates s from l. The *vaporization* curve of the vapor pressure separates l from g and ends in the *critical point* B above which the dissimilarity of liquid and gas vanishes. Indeed, the coexistence of all three phases is possible only in the triple point with unique values T, p. This means, no freedeom is left to choose any of the thermal variables arbitrarily.

(d) For a *solution* of a substance in a solvent $\Gamma = 2$. If we ask for the coexistence with vapor, we obtain $\Phi = 2$, and $f = 2$. Indeed vapor pressure depends on T and on the concentration of the solution. This example will be discussed in chapter 3.1.3.

The eutecticum. The solution shows more different characteristics in the coexistence of the liquid and the solid phase. Let us consider the example of ammonium chloride in water. Γ is always 2. In Figure 15 the *phase diagram* of the system in the liquid and the solid states for a given pressure, say the atmospheric pressure, is roughly schematized. There are different areas in the diagram in which the vertical axis is temperature and the horizontal axis is the relative concentration (mole fraction) x of chloride. We designate water by w and chloride by c if in the liquid phase, and add an asterisk on the letter if the corresponding substance is in the solid state. For sufficient high temperature, both components are in the liquid phase and form together a homogeneous mixture. $\Phi = 1$, and thus $f = 3$. Indeed pressure, temperature and x are free. This means that they can assume values independent from one another. We consider, however, only one value of pressure. If temperature is lowered, we come into a region of the diagram, where one of the two substances crystallizes. The falling out of this substance changes the concentration x of chloride in the remaining solution and leads it to a certain value x_0. This liquid is called the *eutectic* solution or the *eutecticum*. We designate it by e. If originally x was lower than x_0, ice w* will fall out. If x was higher than x_0, solid chloride c* will fall out. In these regions $\Phi = 2$ and thus $f = 2$. As pressure is given, only one degree of freedom remains to be arbitrarily given, the temperature. The concentration of the liquid is then uniquely x_0. In the special case that already from the beginning the solution had the eutectic concentration, the solution freezes to the homogeneous solid state e* like a pure substance at a certain temperature T_f, the freezing point of the eutecticum. If at the beginning the concentration x was different from x_0 below this temperature, in the frozen eutecticum crystals remain, of ice or of chloride.

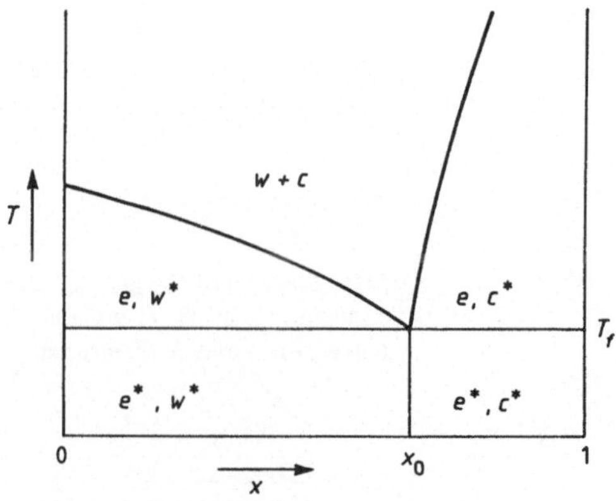

Figure 15

The "eutecticum". Schematical phase diagram of a solution of a substance c of concentration (mole fraction) x in water at different temperatures T. The different homogenous phases are: solution $w + c$, pure ice w^*, solid pure substance c^*, and the eutecticum e in the frozen state. T_f is the freezing temperature of the eutecticum.

2.3.8   Stability Relations

In this chapter we consider a consequence of the convexity of availability as a function of the extensities. Due to this property the absolute thermodynamic equilibrium of a system is stable in the sense that any infinitesimal deviation from the equilibrium raises the tendency to return to the equilibrium. We shall still see another consequence, the so-called *Le Chatelier-Braun* principle.

We return to the convenient notation

$$\Lambda = kT^0 K(\rho, \rho^0) \tag{8.1}$$

for the phenomenological availability with the macroscopic expression

$$K(\rho, \rho^0) = I - I^0 + \lambda^0_\nu (M^\nu - M^{0\,\nu}) \geqslant 0. \tag{8.2}$$

For a small deviation $\delta\rho$ from state ρ, we obtain with the *Gibbs* fundamental equation

$$\frac{\partial I}{\partial M^\nu} = -\lambda_\nu, \tag{8.3}$$

the inequality

$$K(\rho + \delta\rho, \rho) = \frac{1}{2} \frac{\partial^2 I}{\partial M^\mu \, \partial M^\nu} \, \delta M^\mu \delta M^\nu + ... \geqslant 0. \tag{8.4}$$

This expression is equal the second order term in δM of the change

$$\delta K(\rho, \rho^0) = -(\lambda_\nu - \lambda^0_\nu)\delta M^\nu + \frac{1}{2} \frac{\partial^2 I}{\partial M^\mu \, \partial M^\nu} \, \delta M^\mu \delta M^\nu + ... \tag{8.5}$$

of $K(\rho, \rho^0)$ induced by the change $\delta\rho$ of ρ. The inequality thus reads that the availability is a convex function of the extensities M. We can write the inequality (8.4) also in the form

$$-\frac{\partial \lambda_\nu}{\partial M^\sigma} \, \delta M^\nu \delta M^\sigma \geqslant 0, \tag{8.6}$$

and in the still more concise form

$$-\delta\lambda_\nu \delta M^\nu \geqslant 0. \tag{8.7}$$

The *Le Chatelier-Braun* principle. According to the discussion in connection with eq. (7.4), $\lambda_\nu - \lambda^0_\nu$ has the character of a resisting "force" with which the system acts against a deviation of an extensity M^ν from the equilibrium value. This "force" is an expression for the tendency to drive the system back into the equilibrium ρ^0. The change $\delta\lambda_\nu$ of this "force" which is caused by a change $\delta\rho$ of the state ρ occurs in (8.7). – The situation is similar to that in a mechanical parabolic potential

$$\varphi(M) = \frac{\gamma}{2} (M - M^0)^2, \tag{8.8}$$

where M is the *Cartesian* coordinate of a mass point with the equilibrium value M^0. The corresponding restoring force is

$$F = -\frac{\partial \varphi}{\partial M} = -\gamma(M - M^0). \tag{8.9}$$

An increase δM of M yields a change δF of this force satisfying

$$- \delta F \, \delta M = \gamma (\delta M)^2 \geqslant 0 \tag{8.10}$$

in analogy to inequality (8.7). The restoring force becomes stronger if the change of the state goes away from the equilibrium, it becomes weaker if the change goes toward the equilibrium. – In thermodynamics the corresponding statement is the *Le Chatelier-Braun* principle:

> Any change of the thermal state of a system makes the resistance against a deviation from the equilibrium stronger if the resistance works against the change, and makes it weaker if the resistance favors the change.

Therefore the absolute equilibrium state is stable against small fluctuations. The spontaneous development always leads back to the equilibrium.

The sign of susceptibilities. We can write inequality (8.7) in still another form:

$$- \frac{\partial M^\nu}{\partial \lambda_\sigma} \delta \lambda_\nu \delta \lambda_\sigma \geqslant 0. \tag{8.11}$$

The two inequalities (8.6, 8.11) represent the two statements – dependent on each other – that the two susceptibility matrices $\partial \lambda / \partial M$ and $\partial M / \partial \lambda$ possess a negative characteristic quadratic form. We have obtained this result already in the general statistical theory in chapter 1.3.2. We should, however, stress that now in the phenomenological framework the result is a consequence of the Second Law.

In particular we obtain for any ν the inequalities

$$\frac{\partial M^\nu}{\partial \lambda_\nu} \leqslant 0, \qquad \frac{\partial \lambda_\nu}{\partial M^\nu} \leqslant 0. \tag{8.12}$$

With performing the derivative in the first expression, all λ_σ with $\sigma \neq \nu$ are kept constant, in the second, however, all corresponding M^σ are kept constant. Therefore these susceptibilities in general are not reciprocal to each other.

Some typical examples of susceptibilities shall be given. With

$$k \lambda_0 = \frac{1}{T}, \qquad k \lambda_l = \frac{y_l}{T} \tag{8.13}$$

we obtain for instance for each l (which now is not a summation dummy)

$$\left(\frac{\partial M^l}{\partial y_l} \right)_T \leqslant 0. \tag{8.14}$$

Thus in particular the compressibility of a gas or a liquid

$$\kappa = - \frac{1}{V} \left(\frac{\partial V}{\partial p} \right)_T \tag{8.15}$$

is positive. The pressure of a gas increases if it is compressed, thus increasing the resistance against the compression, correspondingly to the *Le Chatelier-Braun* principle. Otherwise the system would be unstable against fluctuations. Analogously, the magnetization M increases with the magnetic field B if the temperature is kept constant:

$$\left(\frac{\partial M}{\partial B}\right)_T \geqslant 0.$$
(8.16)

So far we considered working parameters. But also for particle numbers N of any species we obtain

$$\left(\frac{\partial N}{\partial \mu}\right)_T \geqslant 0.$$
(8.17)

Thus N increases with increasing μ. The tendency of equilibration of μ in space is connected with the tendency of the diffusion in a solution towards homogeneous particle density.

Positivity of heat capacity. To bring also the case that M^ν is energy into the scheme, in which (T, y) are the independent variables, requires a separate consideration. The sign of the quadratic form (8.11) is independent of the size of the $\delta\lambda$. The inequality

$$-\frac{\partial M^\nu}{\partial \lambda_\sigma} \lambda_\nu \lambda_\sigma = \frac{\partial M^\nu}{\partial \lambda_\sigma} \frac{\partial I}{\partial M^\nu} \lambda_\sigma = \frac{\partial I}{\partial \lambda_\sigma} \lambda_\sigma \geqslant 0$$
(8.18)

holds, too. For y_l kept constant we obtain

$$\left(\frac{\partial I}{\partial T}\right)_y = \frac{\partial I}{\partial \lambda_\nu}\left(\frac{\partial \lambda_\nu}{\partial T}\right)_y = -\frac{1}{T}\frac{\partial I}{\partial \lambda_\nu}\lambda_\nu.$$
(8.19)

Entropy S is equal $-kI$, thus

$$T\left(\frac{\partial S}{\partial T}\right)_y = C_y \geqslant 0.$$
(8.20)

If the system is materially closed, in our designation all y_l are the conjugate p_l to working parameters V^l. Then C_p is the "heat capacity" to the heating process in which the parameters p_l are held constant. This is a consequence of the general connection

$$T\,dS = đQ$$
(8.21)

for reversible changes. *Heat capacity* is the heat necessary to increase temperature by one degree centigrade. As (8.18) holds for any set λ, we obtain the positivity of the heat capacity for any set of intensities p_l. Also if the volume is held constant and no other working parameter occurs in (8.18), we obtain

$$T\left(\frac{\partial S}{\partial T}\right)_V = C_V \geqslant 0.$$
(8.22)

With the *Gibbs* fundamental equation

$$T\,dS = dU + p\,dV$$
(8.23)

we can also write

$$C_V = \left(\frac{\partial U}{\partial T}\right)_V \geqslant 0.$$
(8.24)

All these relations, the sign of the susceptibilities and of heat capacity, are consequences of the *Le Chatelier-Braun* principle and express the stability of the thermal equilibrium. Or in a still more general way, they are consequences of the convexity of availability as function of the extensities. It should, however, be stressed that this does not involve that availability is also convex if expressed as a function of the intensities or of other thermal variable sets.

2.3.9 Specific Heat

As explained in the preceding chapter, heat capacity to intensity values p_l kept constant is given by

$$C_p = T \left(\frac{\partial S}{\partial T} \right)_p. \tag{9.1}$$

For a homogeneous substance, the corresponding substantial property is best described by the *specific heat*. This is heat capacity of a standard mass. It is advantageous to choose one mole as standard mass because this brings a certain uniformity for whole classes of substances. We shall designate the molar specific heat by the small letter c.

Also other restrictions for the heating process instead of keeping intensities p_l constant are often important. Such restrictions distinguish a certain direction γ in the space of independent thermal variables. For this direction

$$c_\gamma = T \left(\frac{\partial s}{\partial T} \right)_\gamma \tag{9.2}$$

is generally called a *polytropic* specific heat. If for instance all working parameters V^l are kept constant, then the corresponding specific heat is

$$c_V = T \left(\frac{\partial s}{\partial T} \right)_V. \tag{9.3}$$

With *Gibbs* fundamental equation for one mole of the substance

$$T \, \mathrm{d}s = \mathrm{d}u + p_l \, \mathrm{d}v^l \tag{9.4}$$

we obtain

$$c_V = \left(\frac{\partial u}{\partial T} \right)_V. \tag{9.5}$$

By use of the enthalpy per mole

$$j = u + p_l v^l \tag{9.6}$$

eq. (9.4) assumes the form

$$T \, \mathrm{d}s = \mathrm{d}j - v^l \, \mathrm{d}p_l \tag{9.7}$$

and we obtain

$$c_p = T \left(\frac{\partial s}{\partial T} \right)_p = \left(\frac{\partial j}{\partial T} \right)_p. \tag{9.8}$$

A connection between c_p and c_V results as follows. Equation

$$\left(\frac{\partial s}{\partial T}\right)_V = \left(\frac{\partial s}{\partial T}\right)_p + \left(\frac{\partial s}{\partial p_l}\right)_T \left(\frac{\partial p_l}{\partial T}\right)_V \tag{9.9}$$

and the *Maxwell* relation eq. (5.13) of chapter 2.3.5

$$\left(\frac{\partial s}{\partial p_l}\right)_T = -\left(\frac{\partial v^l}{\partial T}\right)_p \tag{9.10}$$

yield

$$c_p - c_V = T\left(\frac{\partial p_l}{\partial T}\right)_V \left(\frac{\partial v^l}{\partial T}\right)_p. \tag{9.11}$$

This difference depends on the thermal equations of state only, whereas c_V or c_p depend on the caloric equation of state as well. For a gas with volume v as the only working parameter and with pressure p we obtain in particular

$$c_p - c_V = T\left(\frac{\partial p}{\partial T}\right)_V \left(\frac{\partial v}{\partial T}\right)_p. \tag{9.12}$$

The *ideal gas* equation

$$pv = RT \tag{9.13}$$

yields for the ideal gas

$$c_p - c_V = R. \tag{9.14}$$

We return to the general relations (9.8, 9.10) for any gas. They lead us to

$$\left(\frac{\partial c_p}{\partial p_l}\right)_T = -T\left(\frac{\partial^2 v^l}{\partial T^2}\right)_p. \tag{9.15}$$

Eq. (9.5) and the *Maxwell* relation (5.12) of chapter 2.3.5, now written in the form

$$\left(\frac{\partial s}{\partial v^l}\right)_T = \left(\frac{\partial p_l}{\partial T}\right)_V \tag{9.16}$$

lead us analogously to

$$\left(\frac{\partial c_V}{\partial v^l}\right)_T = T\left(\frac{\partial^2 p_l}{\partial T^2}\right)_V. \tag{9.17}$$

In particular, this yields that c_V of the ideal gas is independent of v.

Heat capacity as a statistical measure. Once more we shall leave the phenomenological framework to consider a connection between heat capacity and the general statistical measure *bit-number variance*[12]. The *Shannon* information I is the mean value and thus the first cumulant of the bit-number

$$\mathbf{b} = -\ln \boldsymbol{\rho}. \tag{9.18}$$

---

[12] *F. Schlögl*, Z. Physik **267**, 77 (1974).

For the following it does not matter whether ρ is the probability distribution in classical Γ-space or the statistical operator in quantum mechanics. (Compare eq. (1.9) of chapter 1.2.1 and eq. (4.28) of chapter 2.1.4.) We can find all cumulants of $\mathbf{b}$ with the generating function

$$\Gamma(\alpha) = \ln \operatorname{tr} [\rho \exp(\alpha\mathbf{b})] = \ln \operatorname{tr} \rho^{1-\alpha} \tag{9.19}$$

as explained in chapter 1.1.4. This function has distinguishing features which transfer themselves to the cumulants of any order.

The first one is that each cumulant is additive for uncorrelated subsystems I and II because for the compound system the relations

$$\rho = \rho^{\mathrm{I}}\rho^{\mathrm{II}}, \qquad \mathbf{b} = \mathbf{b}^{\mathrm{I}} + \mathbf{b}^{\mathrm{II}} \tag{9.20}$$

hold. Therefore we can expect that the cumulants of $\mathbf{b}$ depend sensitively on the development of correlations between parts of a system. These can be spatial parts as well as different sets of degrees of freedom. (For instance the carrier electrons in the valence band of a semiconductor on one hand and the lattice band on the other hand are nearly uncorrelated in several respects but not separated in space.)

The second distinguishing feature of the bit-number cumulants is that they are invariant with respect to a permutation of the corresponding sample set. In a space of continuous events, particularly in Γ-space, they are invariant against volume conserving transformations. Therefore they remain unchanged by a *Liouville* motion. That is to say, a change of these cumulants with time indicates that the dynamics in Γ-space is not reversible.

We shall designate the *fluctuations* of a quantity $\mathbf{M}^\nu$ in a distribution ρ with

$$\Delta\mathbf{M}^\nu = \mathbf{M}^\nu - \langle M^\nu \rangle = \mathbf{M}^\nu - M^\nu. \tag{9.21}$$

(This use of the symbol Δ is not to be confused with that of the former chapters for macroscopic differences.) For the generalized canonical distribution

$$\rho = \exp(\Psi - \lambda_\nu \mathbf{M}^\nu), \tag{9.22}$$

the bit-number is

$$\mathbf{b} = -\Psi + \lambda_\nu \mathbf{M}^\nu. \tag{9.23}$$

The second cumulant, that is the *variance* of $\mathbf{b}$, is

$$\langle (\Delta b)^2 \rangle = \lambda_\nu \lambda_\sigma \langle \Delta M^\nu \Delta M^\sigma \rangle. \tag{9.24}$$

On behalf of the connection eqs. (2.23, 2.27) of chapter 1.3.2 between fluctuation correlations and susceptibilities

$$\langle \Delta M^\nu \Delta M^\sigma \rangle = -\frac{\partial M^\nu}{\partial \lambda_\sigma} \tag{9.25}$$

we obtain

$$\langle (\Delta b)^2 \rangle = -\lambda_\nu \lambda_\sigma \frac{\partial M^\nu}{\partial \lambda_\sigma}. \tag{9.26}$$

The right hand side is the expression of eq. (8.18). As $-kI$ is equal to S, corresponding to eq. (8.19, 8.20), it is

$$\langle (\Delta b)^2 \rangle = \frac{1}{k}\, C_y. \tag{9.27}$$

This shows that heat capacity C_p is a special case of the general statistical measure "bit-number variance". Heat capacity C_p to constant intensities p_l is in so far a special case as the corresponding ρ is a generalized canonical distribution in which only working parameters V^l and energy occur as random quantities. We see now in a very direct way that C_y is always positive.

The sensitivity of the measure bit-number variance with respect to interior correlations in a system is realized by the dramatic behavior of specific heat at a critical point of a phase transition due to the building up of "critical correlations" in the substance.

In chapter 2.2.4 we already discussed that the same thermal state can be represented by different distributions ρ depending on which extensities M^ν are introduced as random variables and which ones as sharp parameters. Therefore it should be emphasized that ρ associated with $\mathbf{b}$ in eq. (9.27) is this one in which all M^l conjugate to the y_l kept constant, occur as random variables. For instance, if we replace the pressure ensemble, which leads to specific heat c_p for constant pressure, with the canonical distribution, we obtain the specific heat c_V for constant volume, instead of c_p.

2.4 The Low Temperature Regime

If we progress to low temperatures, the deviations from classical physics become more important and the thermodynamic behavior of a system is increasingly ruled by quantum mechanics. Indeed, the historical development of quantum mechanics was closely connected with low temperature pyhsics.

The basic concepts of the statistical method in general thermodynamics are not changed by quantum mechanics. That is to say, the method of associating probabilities to microstates are not changed. Different, however, are the sets of the microstates. A characteristic property of these states in quantum mechanics is that they are enumerable. In particular the fact that the first excited energy level is separated from the ground state by a finite "gap" influences the low temperature behavior of systems. Another characteristic property of the whole set of microstates of many-particle systems arises from the indistinguishability of particles of the same species. Due to this feature, the sets are considerably changed against classical physics. These changes are fundamentally different for the two different kinds of particles, the *bosons* and the *fermions*. A third difference from classical mechanics is the dynamics of quantum states. A striking example is the tunneling effect which becomes predominant at extremely low temperatures.

Of course the behavior of macroscopic systems is influenced by quantum mechanics not only at low temperatures. For instance, the quantum properties of a fermion gas are of fundamental importance in physics of metals and semiconductors

not only at low temperature. In the following, however, we shall confine the discussion
to more general characteristics of low temperature thermodynamics, less dependent on
more individual properties of special systems.

2.4.1   The *Nernst* Theorem

In this chapter we shall be concerned with the statistical foundation of a theorem
given 1906 by *W. Nernst* on macroscopic arguments. It was then reformulated by
M. Planck as the theorem that entropy of a body of uniform matter goes to zero during
the approach to the absolute zero point of temperature. This fact makes it possible to
calculate thermodynamic quantities which otherwise could not be obtained uniquely,
as unknown integration constants occur. This is of particular importance for applications
of thermodynamics in chemistry. The theorem is also called the *Third Law* of thermodynamics.

In the statistical theory this theorem can be understood as a consequence of
the existence of discrete energy levels in quantum mechanics. In general, these levels
are degenerate so that not only one but say g_n microstates i belong to the same energy
value E_n. The degree g_n of degeneration is very high in a macroscopic system. In the
canonical distribution to a given temperature

$$T = \frac{1}{k\beta} ,$$

$$(1.1)$$

the probability of only one of these microstates is

$$P_i = \exp \beta(F - E_n).$$

$$(1.2)$$

The probability of the energy value E_n is

$$P(E_n) = g_n \exp \beta(F - E_n).$$

$$(1.3)$$

It is convenient to designate the energy levels in agreement with the sequence of their
values

$$E_{n-1} < E_n.$$

$$(1.4)$$

Then E_0 is the ground level. The following equation holds:

$$\frac{P(E_n)}{P(E_0)} = \frac{g_n}{g_0} \exp\left[-\beta(E_n - E_0)\right].$$

$$(1.5)$$

If the temperature T is so low that for the energy gap between the first excited and the
ground level the relation

$$E_1 - E_0 \gg kT$$

$$(1.6)$$

holds, then all ratios eq. (1.5) become so small that practically only the lowest level E_0
with g_0 microstates is occupied. Each of these microstates has the same probability with
the value $1/g_0$. Therefore entropy of the system in the ground level is

$$S = k \ln g_0.$$

$$(1.7)$$

Macroscopically observable entropy values have to be of the order kL if L is the number of molecules in a mole. As already mentioned, this number, which is called the *Loschmidt number*, is of the extremely high order of 10^{23}. We can assume that, therefore, S is extremly small compared with kL. This quantity kL is equal to the gas constant R as we shall see in chapter 4.1.1. Independently of this, however, we already encountered in chapter 2.2.3 the equipartition theorem due to which kL is of the order of a macroscopic heat capactiy. To obtain an estimate for S we assume that the degree g_0 of degeneration of E_0 is

$$g_0 < f^\alpha, \tag{1.8}$$

where f is the number of degrees of freedom of the macroscopic system and α is an adequately adapted exponent. If ν is the number of degrees of freedom of one molecule, then

$$f = \nu L, \tag{1.9}$$

$$S < k\alpha(\ln \nu + \ln L) \approx k\alpha \ln L. \tag{1.10}$$

Now $\ln L$ is somewhere around 50, and α will never be comparable with L. This means that S vanishes on the macrosopic scale; i.e. compared with kL:

$$\lim_{T \to 0} S = 0. \tag{1.11}$$

This is the *Nernst* theorem in *Planck*'s formulation.

The theorem was first postulated in phenomenology on the basis of observations of chemical reactions. The conclusions leading to the theorem were not stringent and rather intuitive. The theorem was often called in question because the conditions for its validity were not realized in many cases. The first condition is the existence of an energy gap of sufficient size to fulfil inequality (1.6) for an observable temperature T. More problematic, however is the condition that, with the approach to the zero point of T, the system really goes into the ground level of lowest energy. In certain cases this never happens because it requires a transition over a too high energy barrier. A standard example for this is the diamond which would not go into the ground state even under extreme cooling. The ground state would be graphite. But nevertheless the diamond is extremely permanent and stable, albeit in a inhibited and not in an absolute thermal equilibrium. More complicated situations can be found in glasses. These are not thoroughly ordered crystals, rather with the amorphous structure of a liquid in a "frozen" configuration.

Notwithstanding these restictions, the theorem became a powerful tool particularly in chemistry because in many cases it allowed entropy to be obtained as a function of the thermal state by caloric measurements. This will be explained in the next chapter.

2.4.2 Characteristic Quantities at Zero Point

The vanishing of macrosopic entropy S with vanishing absolute temperature has the consequence that other important quantities also vanish at the "absolute zero point" of temperature.

Specific heat. For any polytropic specific heat we can write

$$c_\gamma = T \left(\frac{\partial s}{\partial T}\right)_\gamma = \left(\frac{\partial s}{\partial \ln T}\right)_\gamma . \tag{2.1}$$

With approach to the absolute zero point of temperature the entropy s of a body of uniform matter goes to zero, whatever the value of other thermal variables may be. $\ln T$ goes to minus infinity. The slope of the curve s over $\ln T$ also goes to zero. This means that

$$\lim_{T \to 0} c_\gamma = 0. \tag{2.2}$$

Working parameters. The *Maxwell* relation

$$\left(\frac{\partial V^l}{\partial T}\right)_p = -\left(\frac{\partial S}{\partial y_l}\right)_T \tag{2.3}$$

yields by differentiation of the equation

$$\lim_{T \to 0} S(T, y) = 0 \tag{2.4}$$

with respect to y_l:

$$\lim_{T \to 0} \left(\frac{\partial V^l}{\partial T}\right)_p = 0. \tag{2.5}$$

In particular for volume V and magnetization M:

$$\lim_{T \to 0} \left(\frac{\partial V}{\partial T}\right)_p = 0, \tag{2.6}$$

$$\lim_{T \to 0} \left(\frac{\partial M}{\partial T}\right)_B = 0. \tag{2.7}$$

These results specifically involve that any method of measuring temperature by expansion of a thermometric substance breaks down near the absolute zero point. The same is true for any thermometric method based on the temperature dependence of magnetization M or of any other extensity.

The thermal equation of state of an ideal gas is in contradiction to eq. (2.6) because it would yield

$$\left(\frac{\partial v}{\partial T}\right)_p = \frac{R}{p} = \frac{v}{T}. \tag{2.8}$$

This, however, is no objection to the theorem. It is due to the fact that the ideal gas equation is not applicable to arbitrarily low temperatures where no substance behaves like an ideal gas.

Calculation of entropy. The *Nernst* theorem allows us to obtain the entropy of a substance by calorimetric methods. Measuring specific heat c_p for constant intensities p_l as a function of temperature allows us the calculation of

$$s(T, p) = \int_0^T dT'\, c_p(T')/T'. \tag{2.9}$$

Before the theorem was known, an integration constant remained unknown. Therefore in *Helmholtz* and *Gibbs* free energy,

$$F = U - TS, \tag{2.10}$$
$$G = J - TS \tag{2.11}$$

an additive linear function of temperature remained undetermined. Thus various thermodynamic calculations were not allowed which later became possible because of the theorem.

It should be pointed out that the calculation of s with eq. (2.9) requires that c_p is known in detail also at low temperatures. For instance an unobserved phase transition at low temperature may lead to wrong results for $s(T)$ because relative small failures in $c_p(T')$ bring large contributions to the integral of eq. (2.9) due to the factor $1/T'$ in the integrand. It is remarkable that in this way processes at very low temperature are connected by thermodynamics with quantities at room temperature.

Eq. (2.10) shows that at the absolute zero point the *Helmholtz* free energy F and the internal energy U become equal. The "bound energy" TS vanishes. This would be true also if S remained finite. The vanishing of entropy, however, yields that the bound energy does not vanish linearly but in a higher order of T at the absolute zero point. On account of the *Gibbs* fundamental equation the general relations

$$\left(\frac{\partial U}{\partial T}\right)_V = T\left(\frac{\partial S}{\partial T}\right)_V, \tag{2.12}$$

$$\left(\frac{\partial J}{\partial T}\right)_p = T\left(\frac{\partial S}{\partial T}\right)_p \tag{2.13}$$

hold. These quantities are the heat capacities C_V, C_p respectively. Furthermore, generally

$$\left(\frac{\partial F}{\partial T}\right)_V = \left(\frac{\partial G}{\partial T}\right)_p = -S. \tag{2.14}$$

Thus on account of the *Nernst* theorem, all these derivatives vanish at the zero point. U and F become practically equal already in a larger regime of T by approaching the zero point. The corresponding is true for J and G.

Chemical activity and reaction heat. The *Nernst* theorem is valid for a materially uniform body. Also homogeneous mixtures of chemical reactants are of this type. They are compared with each other before and after a chemical reaction. By defining isochoric and isobaric *reaction heat*

$$Q_V = - \Delta U, \tag{2.15}$$

$$Q_p = - \Delta J, \tag{2.16}$$

and the corresponding *activities* of a chemical reaction

$$A_V = - \Delta F, \tag{2.17}$$

$$A_p = - \Delta G, \tag{2.18}$$

the differences Δ of the corresponding thermodynamic quantities are those of the reactants after the reaction minus those before the reaction. The amounts of the reactants are standard mole numbers and the turn-over of the reaction has to be complete. Reaction heat and the corresponding activity A become equal at the zero point but moreover their derivatives with respect to T go to zero there. This is sketched schematically in Figure 16. Practically the two quantities A and Q already become equal in a larger domain of temperature near the zero point. Thus the erreneous opinion, first uttered by *Berthelot*, that reaction heat may be a measure for the tendency of a reaction, becomes assymptotically true at low temperatures.

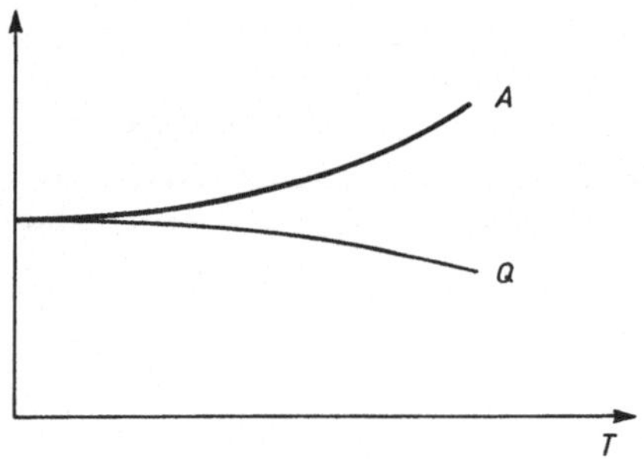

Figure 16

Due to the *Nernst* principle, "reaction heat" Q and "activity" A become practically equal in the neighbourhood of the absolute zero point of temperature.

2.4.3 Unattainability of Absolute Zero Point

We shall show that it is impossible to reach the absolute zero point of temperature by a finite number of cooling steps. A cooling step requires a lowering of entropy S. Things become more transparent if we separate the lowering of temperature from the lowering of entropy in separate steps. It is possible to lower entropy without a change of temperature by an adequate change of working parameters V^l, in the course of which the system is in contact with a heat reservoir. This process is isothermal. Lowering the temperature then is possible by a subsequent adiabatic restoration of the initial values V^l. Instead of V^l also the conjugates p_l can be changed. To illustrate this, let us consider the cooling of a gas. The first process is an isothermal compression of the gas when the piston is in contact with a heat reservoir. The second one is a reversible adiabatic expansion by which the pressure decreases to the initial value. During the first process, T remains unchanged and S is lowered. During the second process, S is unchanged and T

is lowered. — In the general case, we can approximate any cooling process by a sequence of a sufficient number of small steps of the described type.

Let T', p' be the initial values in one step and let the set p' change to p'' at constant temperature T' in the first process. In the second, the adiabatic process, p' is restored and T' is lowered to T''. Then

$$S(T',p') > S(T',p'') = S(T'',p'). \tag{3.1}$$

In

$$\left(\frac{\partial S}{\partial T}\right)_p = \frac{C_p}{T}. \tag{3.2}$$

C_p is positive and not vanishing if T is not zero. The last statement is a most direct consequence of the fact that C_p is a bit-number variance of a generalized canonical distribution, corresponding to chapter 2.3.9. As explained in chapter 2.3.8, the statement can also be understood as consequence of the stability of the equilibrium. According to eq. (3.2), then S is positive for any finite T. The situation is qualitatively illustrated in Figure 17. The most favorable case for the cooling is that the adiabatic step from T' to T'' is reversible. An irreversible step would heighten S and make T'' not so low. This is shown in the figure by a broken line. We come to the result:

> The absolute zero point cannot be reached by a finite number of cooling steps.

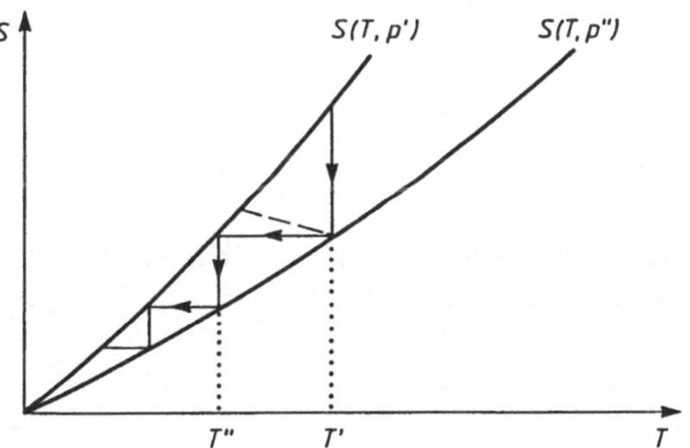

Figure 17

The path of a cooling process with alternating adiabatic and isothermal steps, demonstrating the unattainability of the absolute zero point of temperature.

The *Nernst* theorem is sometimes identified with this *unattainability theorem* to bring it into a form similar to the impossibility principles of certain processes as a formulation of the First and Second Law. It should, however, be stressed that the unattainability theorem is weaker than the *Nernst* theorem because it would also allow S to be finite at the absolute zero point.

*2.4.4   Reactions at Low Temperature

On an atomic level, chemical reactions can only be understood by taking quantum mechanics into consideration. If, however, the binding energies of the reacting molecules are given, thermodynamic calculations are possible for certain characteristic quantities of chemical reactions. But even then some properties of quantum mechanics give rise to changes in comparison to classical physics, in particular in the low temperature regime. The most important ones will be pointed out in this chapter.

The *Arrhenius* factor. Chemical reactions are rearrangements of parts of molecules to new molecules. We consider an elementary system, say for instance a molecule which undergoes a rearrangement of its parts, which we call the "reaction". In general such a rearrangement requires the surmounting of an energy barrier as sketched in Figure 18. There the curve represents the potential energy $\Phi(r)$ depending on a parameter r which characterizes schematically the configuration of the molecule. Let r_0, r_1 be the positions of relative minima of Φ. First we shall discuss the consequences of classical physics. The molecule shall initially be in r_1 and finally in r_0 after the reaction. If $\Phi(r_0)$ is lower than $\Phi(r_1)$ the reaction is exotherm. Energy will be produced by the reaction. Nevertheless, the reaction can occur only if the barrier with the maximum $\Phi(r_m)$ at r_m is surmounted. That means that the *activation energy*

$$\Delta E = \Phi(r_m) - \Phi(r_0) \tag{4.1}$$

has to be conveyed to the reacting molecule. At low temperature this is a rare event. At temperature T we can describe the thermal state in a valley by a canonical distribution over the microstates only which belong to this valley. This corresponds to the knowledge of temperature and the fact that the system is in the valley. The probability of surpassing the barrier by thermal activation therefore is

$$P = C \exp\left(-\frac{\Delta E}{kT}\right). \tag{4.2}$$

This expression is called the *Arrhenius factor* or *activation probability*. So much for the classical theory.

Quantum physics brings essential changes. The first is that the energy levels are discrete. The second is that the ground energy in a valley is higher than the corresponding

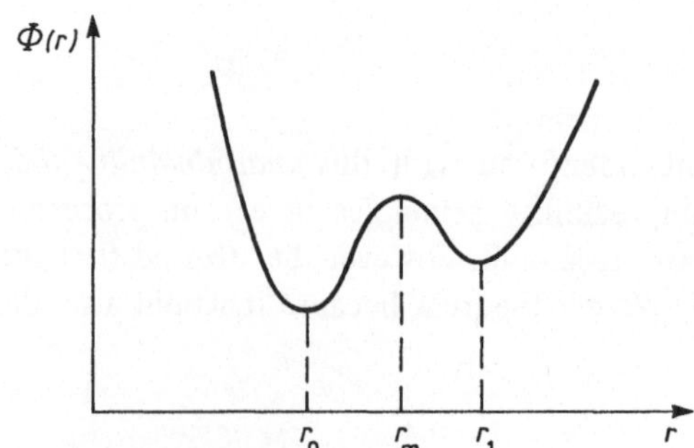

Figure 18

Example of an energy barrier at r_m of the potential energy $\Phi(r)$.

minimum Φ because the quantum mechanical uncertainty relation makes the position r unsharp. We have to put the activation energy equal to

$$\Delta E = E_m - E_1, \tag{4.3}$$

where E_1 is the ground level of the initial valley and E_m is the lowest of its levels higher than $\Phi(r_m)$. This can change the *Arrhenius* factor considerably. As a rule, the reaction probability is enlarged.

Tunneling effect. Quantum mechanics brings still another change. There is a further possibility to surpass the barrier, namely by the tunneling effect which is connected with the wave character of the system [13]. This passing of the barrier occurs without conveyance of energy. If the barrier is thick and high, the tunneling happens very rarely. At extremely low temperatures, however, this remains practically the only way for the reaction and becomes prevalent. For instance, it can be estimated that reactions leading to complex molecules, maybe primitive biological molecules in interstellar clouds, are by a factor of many orders of magnitude more probable than expected without tunneling.

The tunneling effect is important also in solids, in particular in amorphous substances. For instance, it influences transport phenomena also at room temperature.

Once more we should like to stress that these deviations from classical physics are in accordance with the general scheme of thermodynamic statistics which shows how to associate probabilities with microstates. This scheme remains unchanged in quantum physics. What is changed are properties and dynamics of microstates.

---

[13] *V. I. Goldanski,* Annual Rev. of Physical. Chem. **27,** 85 (1976).

3   Macroscopic Description of Special Systems

In the preceding section the general scheme of macroscopic thermodynamics has been developed. In this section this scheme will be applied to special systems. Our intention is not to present a survey over the numerous applications of thermodynamics but to offer some illustrating examples.

First we shall discuss "fluid media", as gases and liquids are called. Because they can be associated with only one working parameter, the volume, they became standard examples in theoretical thermodynamics. Then we shall be concerned with chemical reactions. This is one of the main fields of practical application of macroscopic thermodynamics, and in this field in particular, over a long period the most fruitful ideas in the development of thermodynamics were raised. In the last part of this section we shall demonstrate a specific method of deduction in the application of general thermodynamic principles to special cases. It is the method of cycle processes in so-called "gedanken experiments" (mental experiments), that is characteristic for the particular logical methods in phenomenological thermodynamics we mentioned in the introductory parts to section 2.

3.1   Gases and Solutions

Generally, a chemically pure substance can assume different states of matter, i.e. solid, liquid, or gaseous state. These states are examples of *phases*. The change from one of these states to another one is a *phase transition*. Another example of a phase transition is the transition of a ferromagnet from a magnetic to a nonmagnetic state at a "critical temperature", or the transition of a superconductor from a superconducting state to a normal conducting state.

To describe the change of the thermal state connected with a change of the shape of a solid body, we need many working parameters. The situation is far simpler with gases or liquids. Both are defined as *fluid media*. A gas in a thermal equilibrium will uniformly fill the whole interior of a vessel into which it can enter. Any change of the shape without a change of volume needs no work. Therefore, as a rule, the only working parameter needed in thermal states of a gas is volume. If the amount of a liquid is large enough, we can neglect the effects of surface tension, which, however, are important in small droplets. With this omission the only working parameter of an ideal liquid is volume again. The difference to the gas is the following. A gas offers resistance against any diminishment of the volume by shifting the walls of the vessel. It has the tendency to increase its volume if possible. The liquid offers resistance against any increase as well as any decrease of the volume, once it has assumed an equilibrium volume.

In this subsection we shall not restrict our discussion to pure substances but shall include mixtures of gases as well as liquid solutions. An important fact will be that substances rarefied in a solution behave like rarefied gases as if the molecules of the solvent liquid were not present. Therefore thermodynamics of chemical reactions in dilute solutions becomes very transparent. This will be discussed in detail in the next subsection 3.2.

3.1.1 Ideal Gases and Dilute Solutions

Ideal gas. If gases are sufficiently rarefied and the temperature is not very low, they show a uniform thermal equation of state:

$$p = \frac{RT}{v}, \tag{1.1}$$

as already mentioned in chapter 2.3.5, eq. (5.3). We restate that p is pressure, T temperature, v volume of one mole, and R the gas constant. We also point out that we generally designate extensive variables by a small letter instead of the corresponding capital one if they are related to one mole. The gas constant is the same for all gaseous substances.

In chapter 2.3.5 we saw that the state equation (1.1) implies that internal energy u does not depend on volume:

$$\left(\frac{\partial u}{\partial v}\right)_T = 0. \tag{1.2}$$

Another consequence of eq. (1.1) was obtained in eq. (9.14) of chapter 2.3.9:

$$c_p - c_V = R, \tag{1.3}$$

where c_V, c_p is the molar specific heat for constant volume or constant pressure, respectively. Independent from this is the empirical statement that

$$c_V = \left(\frac{\partial u}{\partial T}\right)_V \tag{1.4}$$

is constant in a large temperature domain defined as *normal domain*. We confine the following discussion to this domain. Then

$$ds = \frac{1}{T}\left(du + p\, dv\right) = c_V \frac{dT}{T} + R \frac{dv}{v} \tag{1.5}$$

yields the entropy per mole

$$s = c_V \ln T + R \ln v + \text{const.} \tag{1.6}$$

With eqs. (1.1, 1.3) we can likewise write:

$$s = c_V \ln p + c_p \ln v + \text{const.} \tag{1.7}$$

The *adiabates* are defined to be the lines of constant value s. In the (p, v)-diagram they are given by

$$p v^\kappa = \text{const}, \tag{1.8}$$

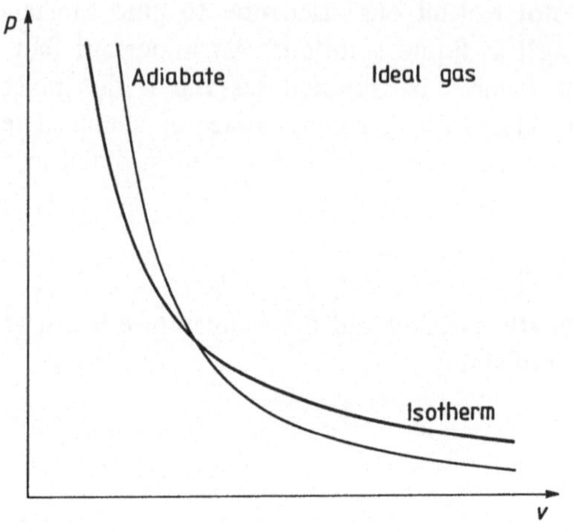

Figure 19

Adiabate (line of constant entropy) and isotherm (line of constant temperature) of the ideal gas in the (p, v)-diagram.

where

$$\kappa = \frac{c_p}{c_V}.$$ (1.9)

Due to eq. (1.3), this exponent is always larger than 1. Therefore, in the (p, v)-diagram the adiabates are decreasing with increasing v with a steeper slope than the isotherms, i.e. than the lines with constant T (see Figure 19).

Remark: As already mentioned in chapter 2.4.2 the ideal equation of state (1.1) is not in agreement with the *Nernst* theorem. The same is true for the entropy of eq. (1.6). The reason is that both equations are valid only in a temperature domain sufficiently far awy from the absolute zero point.

Dilute solutions. The molecules of a dissolved substance in a sufficiently dilute solution behave like the molecules of the rarefied gas which would remain, in an idealized abstraction, if the solvent liquid were removed and the molecules would keep their spatial distribution. The number n of moles of an ideal gas which are in the unit of volume shall be defined as the *mole density*. With it eq. (1.1) yields:

$$p = nRT.$$ (1.10)

The same relation holds for the osmotic pressure p of the dissolved substance. The number n of moles in unit volume then is defined as the *molar concentration* of the dissolved substance. To explain osmotic pressure, we introduce the concept of a *semipermeable wall*. This is a wall through which the solvent but not the dissolved substance can pass. Biological membranes are typical examples. The osmotic pressure acts on the wall from the side of the solution in addition to the normal fluid pressure of the solvent which acts from both sides and therefore gives no resulting force (compare Figure 20).

Figure 21 shows how the osmotic pressure can be measured by the height h of the surface of the solution in a tube which is immersed in the pure solvent and which

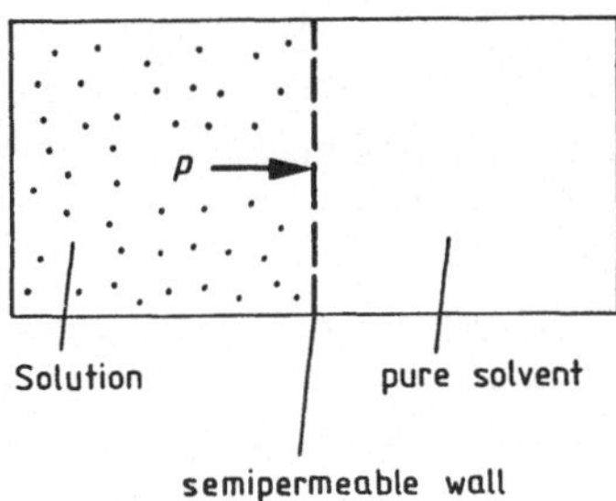

Figure 20
Semipermeable well separating the solution from
the pure solvent for which it is solely permeable.

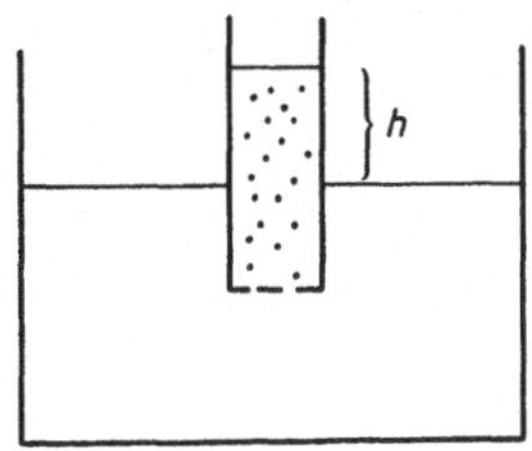

Figure 21
A method of measuring the osmotic pressure with
the rise of the solution over the level of the pure
solvent.

has a bottom that is a semipermeable wall. In the gravitational field, the osmotic pressure is

$$p = \rho g h, \tag{1.11}$$

where g is the gravitational acceleration and ρ the mass density of the solution.

3.1.2 Mixtures of Ideal Gases

If gases are sufficiently rarefied, they do not only obey the ideal equation of state (1.1) but also behave unaffected by one another if they are mixed. This has the *Dalton law* as a consequence. A condition for the validity of this law is that the gases really mix and do not react chemically with one another.

> The *Dalton* law. In a thermal equilibrium of a mixture of ideal gases the total pressure p is equal to the sum of the "partial pressures" p_a of all mixture components.

The partial pressure p_a is the pressure the component a would assume if it were alone in the same volume and at the same temperature as the mixture is. The law reads

$$p = \sum_a p_a. \tag{2.1}$$

(In this context it is advisable to write the summation symbol explicity without using the summation convention associated with dummies.)

Reversible mixing. In general, a mixing process is an irreversible process. We shall, however, show that a certain kind of reversible mixing is thinkable without contradicting thermodynamic principles. We discuss the following "gedanken experiment".

We assume that two ideal gases are in two tubes of equal size which can slide in each other fitting together like two parts of a telescope. This is represented in Figure 22. The ends of the tubes shall be two different semipermeable walls and made so that telescoping in each other is possible. Each wall can be passed only by the gas which was originally in the other tube. The tubes moreover can slide without any friction. Due to the *Dalton* law, the telescoping will be without work at all, because neither of the two gases changes its volume. Thus the mixing and the separation of the gases in these tubes is reversible. The two gases do not influence each other. This is also true for more than two gases. Ideal gases can be mixed and separated reversibly and isothermally if each component does not change the volume it fills out. With that the whole system is isolated. Entropy and internal energy remain unchanged like temperature.

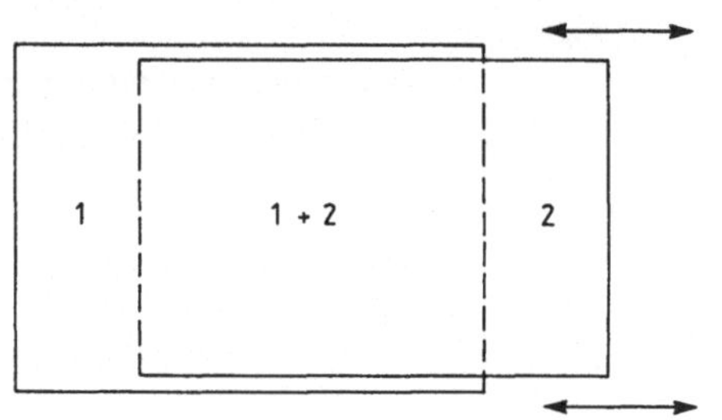

Figure 22

Reversible mixing of two ideal gases by means of two tubes telescoping into each other.

If s_a, u_a are entropy and internal energy per mole of a pure component a of the mixture, and N_a is its mole number, then entropy and internal energy of the mixture are

$$S = \sum_a N_a s_a, \tag{2.2}$$

$$U = \sum_a N_a u_a, \tag{2.3}$$

correspondingly, like for the system of the unmixed components. This involves additivity for the *Helmholtz* free energy analogously:

$$F = \sum_a N_a (u_a + T s_a) = \sum_a N_a f_a. \tag{2.4}$$

The mole volume v_a of the component a is connected with the total volume V of the mixture by

$$V = N_a v_a. \tag{2.5}$$

Therefore we obtain for *Gibbs* free energy:

$$G = F + pV = \sum_a N_a (f_a + p_a v_a) = \sum_a N_a g_a. \tag{2.6}$$

Comparison with the *Gibbs-Duhem* equation for homogeneous systems

$$G = \sum_a N_a \mu_a \tag{2.7}$$

yields

$$\mu_a(T, p, N) = g_a(T, p_a). \tag{2.8}$$

In this equation, the arguments T, p, N of the chemical potential μ_a of a component a in the mixture are the state variables of the mixture. The arguments T, p_a of the molar *Gibbs* free energy g_a, however, are the state variables of the pure component.

Gibbs **paradox.** Let us consider another mixing process which, however, is irreversible. Initially N_1 moles of a gas 1 and N_2 moles of another gas 2 shall be separated in two different parts of a vessel by a wall. Their volumes then are V_1 and V_2, respectively. Now the wall shall be removed. We assume that this can be done practically without any work, for instance by demolishing the wall. The gases can diffuse into each other and will finally be homogeneously mixed and will assume again a thermal equilibrium after a certain time. Then the volume of the mixture is

$$V = \sum_a V_a. \tag{2.9}$$

The whole process shall be performed in thermal contact with a heat reservoir. Thus initial and final temperature of the system are equal. The mixing is isothermal. Entropy before the mixing process was

$$S_{in} = \sum_a N_a s_a(V_a, T) \tag{2.10}$$

and will afterwards be

$$S_{fin} = \sum_a N_a s_a(V, T). \tag{2.11}$$

As temperature remains unchanged, the terms of eq. (1.6) with $\ln T$ cancel each other in the entropy difference ΔS between final and initial state, and we obtain:

$$\Delta S = \sum_a N_a R \ln \frac{V}{V_a}. \tag{2.12}$$

This expression is defined as *mixing entropy*. It is always positive. Also in this mixing process the two components behave independently and do not influence each other. Entropy is increased because each component is expanded from the smaller volume V_a to the larger volume V. As internal energy of the ideal gas is independent of volume, the mixing entropy is determined only by the thermal equation of state and by no other property of the gases.

This implicates a puzzling question. The mixing entropy is the same if the two gases become identical. Then, however, the final and initial state are the same. The occurrence of an entropy change for identical gases therefore is called the *Gibbs paradox*.

A very common argument for solving this paradox is that, due to quantum mechanics, no difference between a property of two gases can be made arbitrarily small. The argument is based on the view that mixing would only occur if the gases are distinguished by different properties and that "self diffusion" is not an irreversible process and therefore not associated with a change of entropy.

In the interpretation of entropy as an information measure, we don't need such an argument, based on special dynamics like quantum mechanics. Entropy depends on which knowledge is taken notice of by the observer. Let the components be two different isotopes of the same chemical element. By special physical methods they are distinguishable, by pure chemical analysis they are not. For two observers who applied different methods the entropy change is different. For one observer the initially separated isotopes were finally mixed. For the other observer, however, the initial and the final state are not distinguishalbe and his knowledge remained unchanged. On the macroscopic level, only the first observer in principle was able to obtain available work by an adequate change from the initial to the final state, e.g. by use of semipermeable movable walls. Thus for the two observers, entropy is different also with respect to the practical consequences. These depend on the knowledge of the observer.

3.1.3   Ideal Mixtures

We have already learned that in a dilute solution the molecules of the dissolved substance behave like a gas as if the solvent liquid were absent. The same is true if not only one but more different substances are dissolved, sufficiently rarefied, and not reacting chemically with one another. The partial pressure of the dissolved substance can be observed as the osmotic pressure. For the osmotic pressure the *Dalton* law holds as well. The dilute solution is a special case of an *ideal mixture* which we shall define on a more general level.

First we return to eq. (2.8), which is satisfied for a mixture of ideal gases. The right hand side of this equation is the molar *Gibbs* free energy of the pure component a if all other components were absent. We find this quantity for the ideal gas through

$$\left(\frac{\partial g_a}{\partial p_a}\right)_T = v_a = \frac{RT}{p_a} \tag{3.1}$$

after the integration with respect to p_a:

$$g_a(T, p_a) = RT \ln p_a + \gamma_a(T), \tag{3.2}$$

where $\gamma_a(T)$ is an "intergration constant" dependent only on T. With the introduction of the independent *mole fractions* of the mixture

$$x_a = \frac{N_a}{N_1 + N_2 + \ldots} \tag{3.3}$$

the *Dalton* law takes the form

$$p_a = x_a p. \tag{3.4}$$

This changes eq. (2.8) to

$$\mu_a(T, p, x) = g_a(T, p) + RT \ln x_a. \tag{3.5}$$

We shall now show that this equation holds also for a larger class of mixtures. These are the above mentioned "ideal mixtures". We define

$$v_a^* = \left(\frac{\partial V}{\partial N_a}\right)_T \tag{3.6}$$

generally as the *partial volume* of the component a in a mixture. This quantity occurs in the homogeneity relation for the volume:

$$V = \sum_a N_a \left(\frac{\partial V}{\partial N_a}\right)_T = \sum_a N_a v_a^*. \tag{3.7}$$

Comparison with the *Gibbs-Duhem* equation

$$G = \sum_a N_a \mu_a \tag{3.8}$$

and

$$\left(\frac{\partial G}{\partial p}\right)_{T, x} = V \tag{3.9}$$

shows that

$$v_a^* = \left[\frac{\partial}{\partial p} \mu_a(T, p, x)\right]_{T, x}. \tag{3.10}$$

In general, this quantity is different from the mole volume

$$(v_a)_p = \left[\frac{\partial}{\partial p} g_a(T, p)\right]_T \tag{3.11}$$

of the pure component a at pressure p. In some mixtures it can even be negative. The "ideal mixture" is now defined by the property that v_a^* and $(v_a)_p$ are equal. This yields

$$\mu_a(T, p, x) = g_a(T, p) + b_a(T, x), \tag{3.12}$$

where b_a is independent of p. This, however, implicates that b_a has to be the same if the mixture is so dilute that the ideal gas relation eq. (3.5) holds:

$$b_a(T, x) = RT \ln x_a. \tag{3.13}$$

Thus eq. (3.5) is always satisfied for any ideal mixture.

Vapor pressure over a solution. We consider a solution of a substance in a liquid and ask for the vapor pressure as a function of the mole fraction x of the dissolved substance. x may be small and thus the mixture ideal. The chemical potential of the solvent is the same μ' in the liquid like μ'' in the vapor if both are in equilibrium at vapor pressure

$$\mu'(T, P, 1 - x) = \mu''(T, P). \tag{3.14}$$

With eqs. (3.12, 3.13) for the solvent, therefore

$$g'(T, P) + RT \ln(1 - x) = g''(T, P).$$ (3.15)

Differentiation of this equation with respect to x yields:

$$\left[\left(\frac{\partial g'}{\partial p}\right)_T - \left(\frac{\partial g''}{\partial p}\right)_T\right]\left(\frac{\partial P}{\partial x}\right)_T - \frac{RT}{1-x} = 0,$$ (3.16)

and with $x \ll 1$:

$$\Delta P = \left(\frac{\partial P}{\partial x}\right)_T x = -\frac{RT}{v'' - v'} x.$$ (3.17)

Thus the vapor pressure of the solution is lower than that of the pure solvent.

$$P(T, x) = P(T, 0) + \Delta P.$$ (3.18)

This yields an increase ΔT of the boiling temperature to a given external pressure p through the equation

$$p = P(T + \Delta T, x) = P(T, 0).$$ (3.19)

As all these changes are small, using the *Clausius-Clapeyron* equation (7.16) of chapter 2.3.7 we can write

$$\Delta P = \left(\frac{\partial P}{\partial x}\right)_{x=0} = -\frac{\partial P(T, 0)}{\partial T} \Delta T = -\frac{q}{v'' - v'} \frac{\Delta T}{T},$$ (3.20)

where q is the heat of evaporation of one mole of the solvent. Comparison with eq. (3.17) yields

$$\frac{\Delta T}{T} = \frac{RT}{q} x.$$ (3.21)

If T is far enough from the critical point, we may neglect v' compared with v''. If we moreover assume the vapor to be an ideal gas

$$v'' = \frac{RT}{P},$$ (3.22)

we obtain the *Raoult law*

$$\frac{\Delta P}{P} = -x.$$ (3.23)

It offers a method of measuring the molecular weight of the dissolved substance.

We can apply eq. (3.17) also to the melting pressure P if we replace v'' with the mole volume of the solvent in the frozen state. Then ΔP is positive for most solvents with the exception of water. In all cases, however, eq. (3.21) holds and yields a lowering of the freezing point. Yet the observations of changes of the freezing point are more difficult than those of the boiling point and require higher values of x.

3.1.4 Real Gases

The behavior of a real gas differs from that of the ideal gas equation (1.1) in particular with increasing concentration of the gas. In Figure 23 a diagram of the equilibrium states of a real gas is sketched, which sould be understood but as a rough scheme to show the qualitative features. In the (pv, p) plane the isotherms are drawn, i.e. the lines of constant temperature T. There is a distinguished temperature T_B, the so-called *Boyle temperature* at which

$$\left[\frac{\partial}{\partial p}(pv)\right]_{p=0} = 0. \tag{4.1}$$

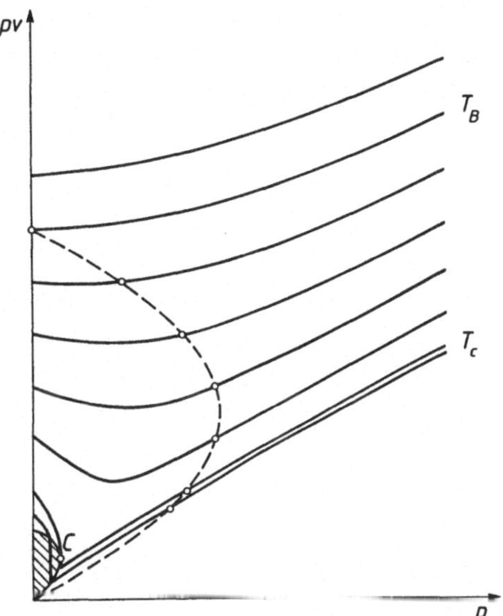

Figure 23

Schematic (pv, p)-diagram of a real gas. The solid lines are isotherms. The isotherm of the *Boyle* temperature T_B runs horizontally into the value for zero pressure. The "ideal points" are connected by a broken line. The coexistence region of gas and liquid is represented as shaded area. The isotherm associated with the critical temperature T_c is tangential to this area at the critical point.

The corresponding isotherm starts horizontally at low pressure. At this temperature therefore a larger domain of pressure p exists in which pv is practically constant, like for the ideal gas. All isotherms to lower temperature pass a minimum of pv. The broken line connects the *ideal points* of the isotherms in which at larger pressure, pv assumes again the value of the ideal gas. Another distinguished temperature is the *critical temperature* T_c. At lower temperature than the critical temperature the gas condensates to a liquid. If the volume is fixed, then, as a rule, only a certain part of the substance becomes liquid, the rest remains in the gaseous state. Both, the liquid phase and the gaseous phase, then are in coexistence. These states of the system correspond to the coexistence regime, which is shaded in the diagram. In this regime, different values of pv are associated with the same value of p, the *vapor pressure*.

The *van der Waals* equation. Of course the described diagram quantitatively becomes quite different for different substances. Nevertheless, a relatively simple schematic model turned out to be very useful. It is the *van der Waals* equation of state:

$$\left(p + \frac{a}{v^2}\right)(v - b) = RT. \tag{4.2}$$

It comprises only two positive parameters a, b to describe individual features of the gas. This is not enough to fit the experimental data and yet the equation became one of the most fruitful relations in the history of thermodynamics.

In comparison with the ideal gas equation (1.1), the pressure p is diminished by a term proportional to v^{-2} corresponding to an attraction between the molecules. Volume v is enlarged by a constant value b corresponding to a finite volume of the molecules.

In Figure 24 the state diagram of the *van der Waals* equation is drawn in a different way than in Figure 23. Four isotherms are drawn in the plane of p and v. One of them is the *critical isotherm* passing the *critical point* C in which

$$\left(\frac{\partial p}{\partial v}\right)_T = 0, \qquad \left(\frac{\partial^2 p}{\partial v^2}\right)_T = 0. \tag{4.3}$$

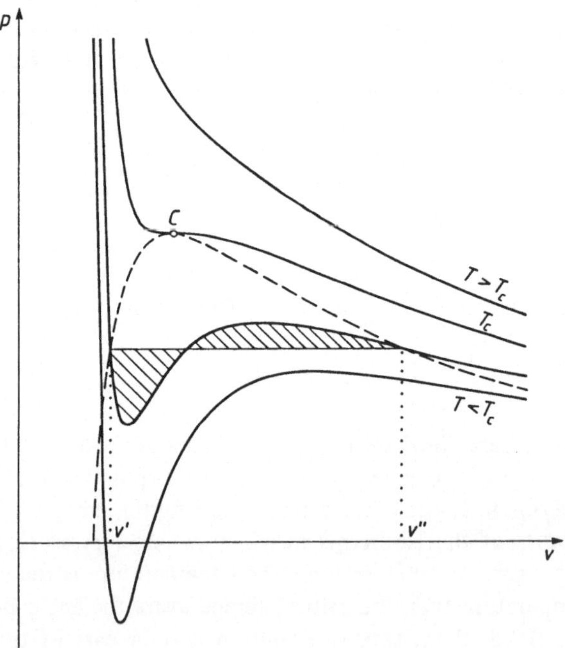

Figure 24 (p, v)-diagram of the *van der Waals* gas. The solid lines are isotherms. The "critical" isotherm of the "critical" temperature T_c passes the "critical point" C. For temperatures below the "critical" one the theoretical isotherms have to be corrected by the *Maxwell* construction due to which the two hatched areas have to be of same size. The region below the broken line is the coexistence regime of gas and liquid. v' is molar volume of the liquid, v'' that of the gas.

This isotherm is associated with the critical temperature

$$T_c = \left(\frac{2}{3}\right)^3 \frac{a}{Rb}\,.$$
(4.4)

Two of isotherms shown in Figure 24, with T lower than T_c, are subcritical isotherms. They pass a relative minimum and maximum. Between these points the *van der Waals* isotherm has a negative slope. This is in contradiction to the stability of equilibrium states. Pressure cannot decrease with increasing volume. Compressibility cannot be negative. Indeed, these isotherms have to be corrected. To obtain agreement with observation, in the interval between the liquid volume v' and the gas volume v'' such an isotherm has to be replaced with a horizontal straight line with constant vapor pressure $P(T)$ depending on T only. In the corresponding states, the volume is partially filled with gas and partially with liquid. – The temperature of the fourth isotherm shown in Figure 24 is higher than T_c. The system there is in a homogeneous state. The discrimination between gas and liquid does not exist any more. The system can be led continuously from the gaseous phase below T_c to the liquid phase with the same temperature, on a path in the (p, v) plane which passes on the upper side round the critical point. According to experimental observation, no discontinuous phase transition occurs in this process.

The *Maxwell* construction. To find the correct value of the vapor pressure $P(T)$, we use the coexistence condition of the two phases, the liquid and the gas state, derived in chapter 2.3.7, eq. (7.13),

$$g'(T, P(T)) = g''(T, P(T)).$$
(4.5)

with the molar *Helmholtz* free energy f, this is identical with

$$f'' - f' + (v'' - v')P = 0.$$
(4.6)

The general relation

$$\left(\frac{\partial f}{\partial v}\right)_T = -p$$
(4.7)

yields:

$$(v'' - v')P = \int_{v'}^{v''} dv\, p.$$
(4.8)

This equation describes the so-called *Maxwell construction* which requires that the shaded areas in the (p, v)-plane between the curvature of the *van der Waals* isotherm and the straight line of constant vapor pressure (the *Maxwell line*) are equal. This prescription determines not only $P(T)$ but also the volumes v', v'' of the two phases. This prescription was derived by *C. Maxwell* by a different argument which required that the reversible work obtained by an isothermal cycle process round the shaded area has to be zero. This argument was, however, vulnerable because it supposed the possibility of leading a process along the unstable states of the curved line. In Figure 24 the broken curved line circumscribes the coexistence regime in which liquid and gas exist simultaneously.

The caloric equation of state. Eq. (5.14) of chapter 2.3.5,

$$\left(\frac{\partial u}{\partial v}\right)_T = -p + T\left(\frac{\partial p}{\partial T}\right)_V \tag{4.9}$$

assumes the form

$$\left(\frac{\partial u}{\partial v}\right)_T = \frac{a}{v^2} \tag{4.10}$$

for a *van der Waals* gas. With the general relation

$$c_V = \left(\frac{\partial u}{\partial T}\right)_V \tag{4.11}$$

we thus obtain

$$\left(\frac{\partial c_V}{\partial v}\right)_T = \frac{\partial^2 u}{\partial T\,\partial v} = 0. \tag{4.12}$$

c_V is not changed even if the gas is rarefied so much that it becomes ideal; hence c_V is the same as for the ideal gas. In the "normal domain" c_V is independent of temperature. Therefore, integrating

$$du = c_V\,dT + \frac{a}{v^2}\,dv \tag{4.13}$$

results

$$u = c_V T - \frac{a}{v} + \text{const.} \tag{4.14}$$

In consequence of eq. (4.14), the gas will be cooled if it irreversibly flows from a smaller into a larger vessel without any work or heat exchange with another system. This is an irreversible adiabatic process without energy change:

$$\Delta u = c_V \Delta T - a\,\Delta(1/v) = 0. \tag{4.15}$$

ΔT is the difference of final minus initial temperature and corresponds to a volume increase associated with the decrease $\Delta(1/v)$ of the mole density. ΔT is always negative and represents a cooling. The reason for this cooling is that the kinetic energy of the gas molecules decreases if their average distance increases. The attractive force between the molecules acts against the expansion. Intramolecular forces of this kind are called *van der Waals forces*.

Cooling by expansion is also possible with an ideal gas if it produces work by moving a piston. Then the kinetic energy of the molecules is lowered and the gas cooled, not by forces between the gas molecules, but by collisions on the moving piston wall.

The *Joule-Thomson* effect. Very different from this is the cooling of a real gas by work against the intermolecular forces in the interior of the gas. It also plays a role in the so called *Joule-Thomson effect*. This effect occurs when a gas is pressed through a porous diaphragm or through a small opening as shown in Figure 25. The piston on the left hand side is pressed against the diaphragm D with pressure p_1. The gas can pass the diaphragm only very slowly and moves the piston on the right hand side against the pressure p_2 which is slightly less than p_1. Because it passes so slowly, the gas practically assumes a thermal state on each side with homogeneous density and homogeneous pressure. If one mole has passed, the whole work done on the gas is

$$W = p_1 v_1 - p_2 v_2 . \tag{4.16}$$

The whole system shall adiabatically be isolated. Therefore the increase of internal energy of the gas is

$$u_2 - u_1 = W. \tag{4.17}$$

This means that the enthalpy

$$j = u + pv \tag{4.18}$$

remains unchanged. For simplicity let us discuss an infinitesimal change:

$$dj = c_p \, dT + \left(\frac{\partial j}{\partial p} \right)_T dp = 0. \tag{4.19}$$

With the so-called *Joule-Thomson coefficient*

$$\tau = - \frac{1}{c_p} \left(\frac{\partial j}{\partial p} \right)_T \tag{4.20}$$

we write

$$dT = \tau \, dp. \tag{4.21}$$

As dp is negative, cooling takes place if τ is positive. In the case we are now discussing, eq. (5.11) of chapter 2.3.5 takes the form

$$\left(\frac{\partial j}{\partial p} \right)_T = v - T \left(\frac{\partial v}{\partial T} \right)_p . \tag{4.22}$$

τ can have different signs in different thermal states. It changes sign if

$$\left(\frac{\partial T}{\partial v} \right)_p = \frac{T}{v} . \tag{4.23}$$

We shall consider the *van der Waals* gas. Let us assume that v is large compared with b when eq. (4.23) is satisfied. The result will justify this assumption. For large v the *van der Waals* equation may then be written as

$$RT = pv + \frac{a}{v} - pb, \tag{4.24}$$

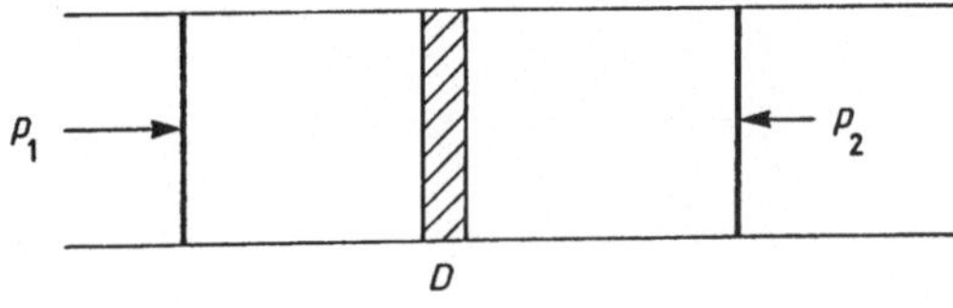

Figure 25

To the *Joule-Thomson* effect. The "throttled" expansion of a gas. The piston on the left hand side is pressed against the diaphragm D with pressure p_1. After passing D, the gas has the lower pressure p_2.

and thus

$$R \left(\frac{\partial T}{\partial v} \right)_p = p - \frac{a}{v^2} .$$

(4.25)

Eq. (4.23) then is satisfied at the so-called *inversion temperature*

$$T_{inv} = \frac{2a}{Rb} = \frac{27}{4} T_c = 6.75 \, T_c.$$

(4.26)

To obtain cooling by the *Joule-Thomson* effect, it is necessary to bring the gas first below the inversion temperature. This effect is the basis of the *Linde* technique of cooling by the repeated "throttled expansion" of a gas.

3.2   Chemical Reactions

It has already been stressed that chemistry is one of the most important fields of application of phenomenological thermodynamics, and corresponding relations have already been discussed. In the following part of this book we shall go into more detail. The first chapter will be concerned with consequences of the First Law, that is to say with the energy balance of a chemical reaction. The central concept will be the *reaction heat* under different conditions. In the second chapter we shall be concerned primarily with consequences of the Second Law when discussing the concept *activity*, which is a measure of the tendency of a reaction. The concepts of reaction heat and activity have already been developed in chapter 2.3.6. Now, however, we shall seek expressions for these quantities which allow us to calculate them accordingly to the special structure of reaction equations. There we shall speak of quantities which are associated with a complete turn-over of reactions. In reality, reactions do not occur completely. To each reaction there is a reverse reaction. Both are competing with each other, striving for an equilibrium in the form of a *chemical equilibrium*. This is an absolute thermodynamic equilibrium without any inhibition of the reactions. In it the initial and final partners of the reactions are present in a certain proportion, as determined by the *law of mass action*. This will be subject of the third chapter. In the last chapter we shall consider the generalization of chemical potentials to the situation of external forces acting on the system. This will lead us to the concept of *electrochemical potentials*.

3.2.1   Reaction Heat

First we shall introduce some designations which allow us to write all reaction equations in a uniform short form. As an example let us choose a very famous reaction, the *Haber-Bosch* synthesis of ammonia which was systematically developed during the First World War when Germany was cut off from natural ammonia resources. The corresponding reaction equation is

$$N_2 + 3H_2 \rightarrow 2NH_3 .$$

(1.1)

N_2, H_2 are the reactants before the reaction, the "initial" reactants. NH_3 is the product of the reaction, the "final" reactant. We shall write such a reaction equation in the general form

$$\sum_a \alpha_a R_a = 0. \tag{1.2}$$

R_a are the symbols for the reactants, α_a are the so-called *stoichiometric coefficients*. We define them positive for the final, and negative for the initial reactants. In our example:

$$R_1 = N_2, \qquad R_2 = H_2, \qquad R_3 = NH_3, \tag{1.3}$$

$$\alpha_1 = -1, \qquad \alpha_2 = -3, \qquad \alpha_3 = +2. \tag{1.4}$$

The reaction equation can be transformed into a balance equation of energy. For this purpose we designate the molar internal energy of the reactant R_a by u_a and write for the increase of internal energy of the reactants by the reaction

$$\Delta U = \sum_a \alpha_a u_a = -Q_V. \tag{1.5}$$

Q_V is the "reaction heat" produced by the reaction. If it is positive, the reaction is "exothermal"; if negative, the reaction is "endothermal", as already defined in chapter 2.3.6. There it was also pointed out that $-\Delta U$ is the isochoric reaction heat Q_V to constant volume. It is associated with the reaction in a *Berthelot* bomb. This, however, is not a necessary condition. It is sufficient that initial and final state are thermal states to the same volume and the same temperature, whichever the intermediate states are. This is a consequence of the First Law.

In an analogous way we obtain the isobaric reaction heat Q_p to equal pressure and temperature in the initial and final state if we replace molar internal energy u_a with molar enthalpy j_a:

$$\Delta J = \sum_a \alpha_a j_a = -Q_p. \tag{1.6}$$

Once more we stress that it is possible that Q_V and Q_p have different signs. Thus the same reaction can be exothermal for constant volume and endothermal for constant pressure or vice versa. So far we have spoken of only one reaction. Often we are concerned with more than one reaction occurring in the same mixture of reactants. To distinguish between them, we shall introduce the additional subscript s to the stoichiometric coefficients and the reaction heat:

$$\Delta J_s = \sum_a \alpha_{as} j_a = -Q_{ps}. \tag{1.7}$$

The First Law makes it possible to determine the reaction heat say of an unobservable reaction, if round about reactions are realizable. The enthalpy difference ΔJ depends only on the initial and final state. The sum of the enthalpy differences of the intermediate reactions is also ΔJ. Care has to be taken if the same reactants are involved in different phases, say solid and gaseous ones. Then the phase transition has to be included into the balance as a reaction with the transition enthalpy as reaction heat. A corresponding extension is of course necessary for ΔU.

Dependence of reaction heat on temperature. Specific heat per mole at constant volume of a substance is

$$c_V = \left(\frac{\partial u}{\partial T}\right)_V . \tag{1.8}$$

If, however, taken at constant pressure, it is

$$c_p = \left(\frac{\partial j}{\partial T}\right)_p . \tag{1.9}$$

In a mixture of ideal gases, energy and enthalpy add independently and we obtain

$$\left(\frac{\partial Q_V}{\partial T}\right)_V = -\sum_a \alpha_a c_{Va}, \tag{1.10}$$

$$\left(\frac{\partial Q_p}{\partial T}\right)_p = -\sum_a \alpha_a c_{pa}. \tag{1.11}$$

These quantities can thus be gained by pure caloric measurements.

3.2.2 Chemical Affinity

We have already pointed out that the adequate measure of the tendency of a reaction to occur spontaneously is the activity and not the reaction heat, as was assumed in the early days of thermodynamics. The more general concept, however, is availability rather than activity. Activity is associated with a chemical reaction and is the change of availability only in the more special cases that the volume or the pressure is held fixed. Availability is distinguished by the Second Law. This leads us to an influence of the Second Law on chemical reactions: the tendency of the reaction to occur if all inhibitions were removed is not determined by the energy change but by the maximum available work.

We are going to discuss isobaric isothermal chemical reactions in which the initial and the final state have the same pressure and the same temperature. Again it does not matter what the intermediate states are. Generally, the maximum available work which can be gained by the reaction is the difference of the availability in the initial state minus that of the final state. In our case, it is the decrease of *Gibbs* free energy associated with the reaction, which simplifies things. Availability depends not only on the momentary state of the system but has to be related to an equilibrium. Thus it depends on two thermal states; in general, without being the difference of two one-state functions. *Gibbs* free energy, however, depends only on one state of the system. For the definition of affinity we do not need the relation to an equilibrium.

Beginning with a mixture of all reactants, of the initial and the final partners, we will discuss a reaction occurring in the mixture with only a very small turn-over.

This means that the difference ΔN_a between the final and the initial values of the mole number N_a of any reactant will be so small that the chemical potentials μ_a will remain practically unchanged. Due to the *Gibbs-Duhem* equation the increase of *Gibbs* free energy in the reaction is

$$\Delta G = \sum_a \mu_a \, \Delta N_a. \tag{2.1}$$

The changes ΔN_a are not independent of one another if the system is materially closed because they are generated only by the chemical reactions. We are excluding exchange of matter with other systems.

It is useful to introduce a reaction rate number $\Delta \xi$. Its definition is, that ΔN_a is equal to $\alpha_a \, \Delta \xi$. If several reactions occur simultaneously, they shall be marked by the subscript s. Then the change of the mole number N_a of species R_a is

$$\Delta N_a = \sum_s \alpha_{as} \, \Delta \xi_s. \tag{2.2}$$

The reaction rate numbers $\Delta \xi_s$ are indepent of one another, contrary to the ΔN_a. In

$$\Delta G = - \sum_s A_s \, \Delta \xi_s \tag{2.3}$$

the quantity

$$A_s = - \sum_a \alpha_{as} \, \mu_a \tag{2.4}$$

is called the *affinity* of the reaction s. The affinity is that linear combination of chemical potentials which really influences the reaction. It is, in metaphorical sense, a kind of driving force for the reaction s.

If ΔG is zero, the chemical equilibrium with respect to all reactions is reached. If the affinity of only one reaction s is zero, then only a momentary equilibration with respect to this particular reaction is reached. In general it will be destroyed immediately by other reactions. In the chemical equilibrium all affinities are zero; for each s the equation

$$\sum_a \alpha_{as} \, \mu_a^0 = 0 \tag{2.5}$$

holds, where each μ_a^0 is the equilibrium value of μ_a. In this equilibrium not only the final but also the initial reactants are present.

The *Le Chatelier-Braun* principle in chemistry. We can interprete the independent $\Delta \xi_s$ as changes of independent extensities M^ν and identify the corresponding affinities with the conjugate intensities $kT\lambda_\nu$. Then for small changes $\delta \xi_s$ of $\Delta \xi_s$ the inequality (8.7) of chapter 2.3.8 yields the *Le Chatelier-Braun* principle in the form

$$\delta \xi_s \, \delta A_s \leqslant 0. \tag{2.6}$$

Whereas in eq. (2.3) we could assume A_s to be unchanged by a sufficiently small $\delta\xi_s$, we are now concerned with the change of A_s in the second order of $\delta\xi_s$. Thus it follows that $\delta\xi_s$ is linked to a change of the numbers N_a and therefore to a change of the chemical potentials μ_a and of the affinities. The affinity A_s is in a figurative sense a "force" restoring the equilibrium and thus a resistance against a deviation from the equilibrium. In accordance with the explanation in chapter 2.3.8, the principle states that this resistance does not only work against a deviation from the equilibrium, but that it moreover increases with the deviation. Here the deviation is a change in the conditions for the reaction which are given with the chemical potentials and finally with the concentrations of the reactants. In the next chapter we shall see that the corresponding also is true for the resistance against another kind of a change of conditions, namely the change of temperature. A special consequence of the principle is the stability of the chemical equilibrium in the presence of small fluctuations. Once more we should like to point out that also this principle is a consequence of the positivity of availability and finally of the positivitiy of information gain.

3.2.3   The Law of Mass Action

The following discussion is restricted to chemical reactions between ideal gases or between substances dissolved in dilute solutions. We know that the molecules in the dilute solution behave like an ideal gas which in an idealization would remain if the solvent were removed and the dissolved molecules would keep their spatial distribution. We have to identify the partial pressure p_a of a gas with the osmotic pressure of the dissolved substance. Then in both cases we can write for the chemical potential of the substance in the ideal mixture

$$\mu_a(T, p, N) = g_a(T, p_a). \tag{3.1}$$

The right hand side of this equation is the molar *Gibbs* free energy which the pure substance R_a would assume if all other mixture components were absent. The ideal gas equation yields

$$\left(\frac{\partial g}{\partial p}\right)_T = v = \frac{RT}{p}, \tag{3.2}$$

$$g(T, p) = RT \ln p + \gamma(T). \tag{3.3}$$

In comparison with eq. (3.2) in chapter 3.1.3, the occurrence of the temperature dependent "integration constant" $\gamma(T)$ now will matter because we shall be interested in the temperature dependence of certain quantities. In chapter 3.1.3 we were only comparing states with the same temperature and therefore this temperature function did not enter into the final results.

First we consider the case that only one reaction can occur. We designate all quantities associated with the chemical equilibrium by the superscript zero. Affinity vanishes in this equilibrium and we may write with the chemical potentials μ_a of the chemical equilibrium

$$A = -\sum_a \alpha_a(\mu_a - \mu_a^0) \tag{3.4}$$

for the affinity in any state which has the same total pressure and the same temperature as the equilibrium. Due to eq. (3.3) thus:

$$A = RT(\ln \Gamma - \ln \Gamma^0), \tag{3.5}$$

where

$$\ln \Gamma = -\sum_a \alpha_a \ln p_a, \tag{3.6}$$

$$\mu_a^0 = RT \ln p_a^0 + \gamma_a(T). \tag{3.7}$$

The mass action constant. The vanishing of the equilibrium value A^0 is expressed by eq. (2.5). Therefore

$$\ln \Gamma^0 = \frac{1}{RT} \sum_a \alpha_a \gamma_a(T). \tag{3.8}$$

This means that the equilibrium value Γ^0 is a function of temperature only. We have thus obtained the following important result:

> At the same temperature, different chemical equilibria are possible with different values of the partial pressures p_a, but they all have the same value of Γ^0.

This law is called the *mass action law* and can be expressed by the equation

$$-\sum_a \alpha_a \ln p_a^0 = \ln \Gamma^0(T). \tag{3.9}$$

The quantity Γ^0 is called the *mass action constant* of the reaction.

Temperature dependence of the chemical equilibrium. With eq. (3.8) we obtain

$$\frac{d}{dT} \ln \Gamma^0 = \sum_a \alpha_a \frac{d}{dT} \left(\frac{\gamma_a}{T} \right) \tag{3.10}$$

and with eq. (3.3), for any reactant,

$$\frac{d}{dT} \left(\frac{\gamma}{T} \right) = \left(\frac{\partial}{\partial T} \frac{g}{T} \right)_p = \frac{1}{T} \left(\frac{\partial g}{\partial T} \right)_p - \frac{g}{T^2} . \tag{3.11}$$

With

$$\left(\frac{\partial g}{\partial T} \right)_p = -s \tag{3.12}$$

and with molar enthalpy

$$j = g + Ts \tag{3.13}$$

the following holds:

$$\frac{d}{dT} \ln \Gamma^0 = \frac{-1}{RT^2} \sum_a \alpha_a j_a . \tag{3.14}$$

The sum in this expression is, in accordance with eq. (1.6), the isobaric reaction heat Q_p. Thus

$$\frac{\mathrm{d}}{\mathrm{d}T} \ln \Gamma^0 = \frac{Q_p}{RT^2} \tag{3.15}$$

As a rule, in a relatively large domain of temperature the reaction heat Q_p does not change very much and can be put approximately constant. We then obtain

$$\Gamma^0 = C \exp\left(-\frac{Q_p}{RT}\right). \tag{3.16}$$

The similarity with the approximation for the vapor pressure eq. (7.19) of chapter 2.3.7 is striking.

We return to the *Haber-Bosch* synthesis of ammonia eq. (1.1) as an illustration. Using the notation of eqs. (1.3, 1.4), we write eq. (3.9) as

$$\Gamma^0 = p_1^0 (p_2^0)^3 (p_3^0)^{-2}. \tag{3.17}$$

Moreover we introduce the relative mole fractions

$$x_a = \frac{N_a}{\displaystyle\sum_b N_b}, \tag{3.18}$$

with which

$$p_a^0 = x_a p, \tag{3.19}$$

and obtain

$$x_1 x_2^3 x_3^{-2} = p^{-2} \Gamma^0. \tag{3.20}$$

The method of *F. Haber* and *R. Bosch* consists in establishing a chemical equilibrium with a high value x_3 of ammonia. The initial reactants hydrogen and nitrogen are present in the air. In principle such an equilibrium can be found at low temperature where Γ^0 is small. There, however, the reaction comes off too slowly. The temperature therefore was taken high, about 550 °C. Then x_3 could be made large by using high pressure p. Indeed, for this method pressure values about 220 atm were used.

Competition of two reactions. We shall again consider an example, namely the two competing reactions of the oxidation of carbon C to carbon dioxide CO_2 or to carbon monoxide CO. We write the reactions directly as balance equations of the enthalpy using the chemical symbols for the molar enthalpies

$$\text{Reaction } s = 1: \quad CO_2 - C - O_2 = \Delta J_1, \tag{3.21}$$

$$\text{Reaction } s = 2: \quad CO - C - \frac{1}{2} O_2 = \Delta J_2. \tag{3.22}$$

Because the two reactions could not be separated from each other, at temperatures below 1000 °C, reaction heat ΔJ_2 was not observed directly. It was, however, obtained indirectly by the round about reaction:

$$\text{Reaction } s = 3: \quad CO_2 - CO - \frac{1}{2} O_2 = \Delta J_3. \tag{3.23}$$

The First Law yields for the isobaric reaction heats $-\Delta J_s$

$$Q_{p2} = Q_{p1} - Q_{p3}. \tag{3.24}$$

All these reactions are exothermal and thus Q_{p1} is larger than Q_{p2}. In the approximation of eq. (3.16)

$$\ln \Gamma_s^0 = -\frac{Q_{ps}}{RT} + \ln C_2, \tag{3.25}$$

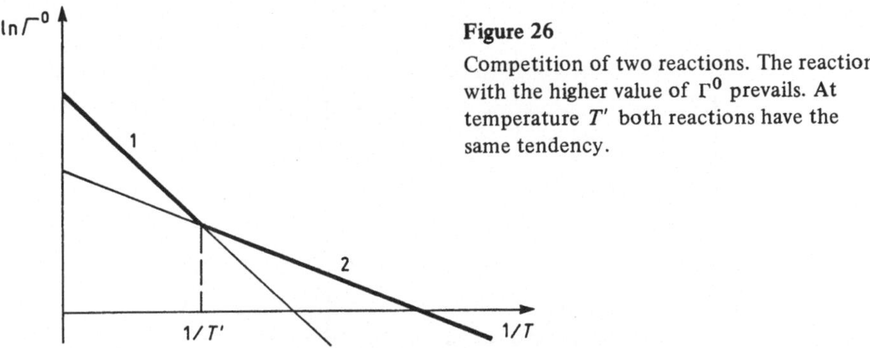

Figure 26

Competition of two reactions. The reaction with the higher value of Γ^0 prevails. At temperature T' both reactions have the same tendency.

we therefore obtain two straight lines with different slopes in the diagram of Figure 26 where $\ln \Gamma_s^0$ is plotted over $1/T$. The two lines cross at temperature T'. The ratio

$$\frac{\Gamma_2^0}{\Gamma_1^0} = \frac{p^0(CO_2)}{p^0(CO)\,[p^0(O_2)]^{1/2}} \tag{3.26}$$

is more than 1 if T is lower than T', and it is less than 1 if T is higher than T'. Room temperature is lower than T' and the oxidation to CO_2 prevails.

Stability of temperature in the chemical equilibrium. To illustrate the *Le Chatelier-Braun* principle in more detail, we consider the dependence of the affinity

$$A = RT(\ln \Gamma - \ln \Gamma^0) \tag{3.27}$$

on temperature and use the approximation of eq. (3.16):

$$\ln \Gamma^0 = -\frac{Q_p}{RT} + \ln C. \tag{3.28}$$

In Figure 27, in the same way as in Figure 26, $\ln \Gamma^0$ is plotted over $1/T$. The straight line with negative slope in case (a) corresponds to an exothermal reaction; the line with positive slope in case (b) corresponds to an endothermal reaction. We consider a mixture of all reactants with given values p_a and thus with a given value Γ of eq. (3.6). The affinity eq. (3.27) vanishes if the chemical equilibrium at temperature T^0 is reached.

If A is positive, the reaction will proceed spontaneously; if A is negative, the reverse reaction will proceed spontaneously. Let us discuss the exotherm reaction (a). If we disturb the chemical equilibrium by lowering the temperature, then A becomes positive, the reaction produces heat and thus works against the disturbance. In case (b) of the endotherm reaction, the reverse reaction will do the same. It is easy to see that for all disturbances the *Le Chatelier-Braun* principle is also satisfied in the following statement:

> Any deviation from the equilibrium temperature changes the affinity in such a way that the reaction heat diminishes the deviation.

Otherwise, the chemical equilibrium would not be stable.

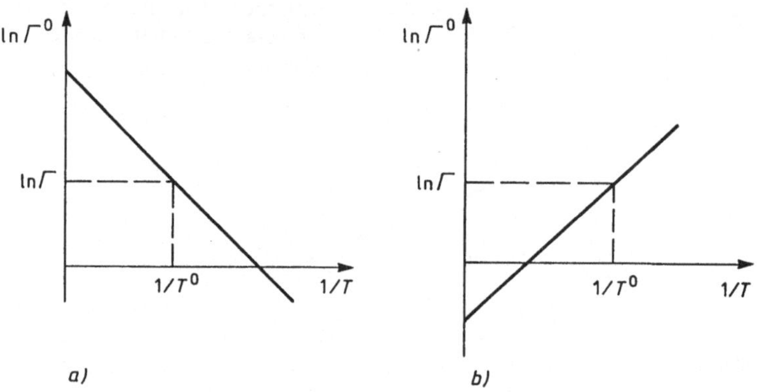

Figure 27 Comparison of an exothermal (a) with an endothermal (b) reaction

3.2.4   Electro-Chemical Potentials

The presence of external forces acting on a system requires a special consideration. Such a situation is at hand if particles of a species a in a mixture possess an electrical charge e_a and if they are being acted upon by an external electrostatic potential $\varphi(\vec{x})$ dependent on coordinates $\vec{x}$ in ordinary space. If $n_a(\vec{x})$ is the spatial density of moles of species a, an infinitesimal change δn_a of these densities is connected with the work

$$\delta W_e = \int d^3x\, \varphi(\vec{x}) \sum_a e_a\, \delta n_a(\vec{x}) \tag{4.1}$$

performed on the system. The condition of thermal equilibrium at fixed temperature T and pressure p is the minimum of *Gibbs* free energy

$$G = U - TS + pV. \tag{4.2}$$

In the *Gibbs* fundamental equation we shall designate an infinitesimal change by the symbol δ instead of the differential symbol d to distinguish it from the volume element d^3x in the integral of eq. (4.1). The *Gibbs* fundamental equation then is

$$\delta U - T\delta S = -p\,\delta V + \delta W_e + \int d^3x\, \mu_a\, \delta n_a. \tag{4.3}$$

For fixed T, p the equilibrium is distinguished by the minimum of *Gibbs* free energy

$$\delta G = \int d^3x \sum_a (\mu_a + e_a\,\varphi)\,\delta n_a = 0. \tag{4.4}$$

If no chemical reactions can occur in the system, the total number of moles of each species is constant. Thus the restriction holds that

$$\delta N_a = \int d^3x\, \delta n_a = 0. \tag{4.5}$$

With *Lagrange* parameters η_a independent of $\vec{x}$, we obtain

$$\mu_a(\vec{x}) + e_a\,\varphi(\vec{x}) = \eta_a. \tag{4.6}$$

These intensities η_a are called *electro-chemical potentials*. In thermodynamic equilibrium they are constant in space in the whole system, whereas the chemical potentials μ_a in general are not

The barometric pressure formula. As an example we consider the isothermal equilibrium of an ideal gas in the gravitational potential $\Phi(\vec{x})$. The mass m acts the part of the electrical charge e_a in eq. (4.6). The chemical potential of the pure ideal gas is according to eq. (3.3)

$$\mu = RT \ln p + \gamma(T). \tag{4.7}$$

With T the term γ is also constant in space. Therefore eq. (4.6) in the form

$$\mu + m\Phi = \text{const} \tag{4.8}$$

yields the so-called *barometric pressure formula*

$$p = p_0 \exp\left(-\frac{m\Phi(\vec{x})}{RT}\right). \tag{4.9}$$

p_0 is the pressure on the surface where Φ is zero, the "zero level". In the homogenous gravitational field with the gravitational acceleration g the formula reads

$$p = p_0 \exp\left(-\frac{mgz}{RT}\right), \tag{4.10}$$

where z is the height above the zero level

At first sight, the dependence of μ_a on space could seem contradictory to the principle of homogeneous intensities in equilibrium. The general principle, however, requires that the intensity λ_ν is equal in all systems which are in unrestricted equilibrium with one another if it is associated with the same independent conjugate extensity M^ν in all these systems. In our case, the extensity conjugate to $\mu_a(\vec{x})$ is the local quantity

$n_a(\vec{x})$ for which the index ν is replaced with the local set $(a, \vec{x})$. These extensities are dependent on one another according to eq. (4.5). Yet the "electro-chemical potentials" η_a are the intensities conjugate to the total number N_a of moles of species a as extensities which are global quantities, not local ones. Thus each one is the same extensity in all parts of the system. Correspondingly η_a is constant in space, whereas $\mu_a(\vec{x})$ is not.

3.3   The Method of Cycle Processes

The First and Second Law of thermodynamics entered into many of the previous deductions in the form of the uniqueness of the state functions "internal energy" and "entropy". These again led us to the existence of different derived state functions, like "free energy" or other thermodynamic potentials. The use of the very existence of such state functions is undoubtedly a very fruitful and direct tool for the deduction of special thermodynamic connections. In the existence of these functions the Basic Laws of thermodynamics are summerized in a very concise but abstract way. A purely formal use, however, can veil the connection with the content of these principles, in particular of the Second Law.

There is another method to apply the Basic Laws, leading to the same results. To be shure, this method sometimes looks a bit laborious but shows clearly the consequences of the Basic Laws in each step of the process. In particular it is the application of the maximum efficiency principle to adequately chosen *Carnot* cycles. In the following we shall demonstrate this method in three examples. The reader who is interested in more examples can find them for instance in the famous book "Wärmelehre" ("Theory of Heat") by *R. Becker*[14]). The method is suited to make thermodynamic considerations more colorful and to develop a practical feeling for thermodynamic connections. As these reversible cycle processes are idealized intellectual constructions not performed in reality, they are often called "gedanken experiments".

3.3.1   An Isothermal Cycle

We restate that the efficiency η of any heat-work engine is the ratio of the work gained by the cycle process to the heat withdrawn from the warmer reservoir. The Second Law in the form of the principle of maximum efficiency states that all reversible heat-work engines possess the same efficiency

$$\eta_r = \frac{T_2 - T_1}{T_2} ,$$
(1.1)

if they work between heat reservoirs of temperatures $T_1 < T_2$.

If in particular the cycle is isothermal, then T_1 is equal T_2 and η_r is zero. If η_r were different from zero, we could withdraw heat from one reservoir and gain work.

---

[14]) *R. Becker*, Theorie der Wärme (Springer, Berlin, Göttingen, Heidelberg 1955);
      R. Becker, Theory of Heat, ed. by *G. Leibfried* (Springer, Berlin, Heidelberg, New York 1967).

The engine would be a perpetuo mobile of second kind, and thus it is impossible. The obtainable work has to be zero. − In the more abstract scheme of thermodynamic potentials this would be a direct result of the existence of the state function *Helmholtz free energy* which remains unchanged after the isothermal cycle, like any unique state function.

The barometric pressure formula. We shall derive once more this formula which was obtained already in chapter 3.2.4. But now we shall use the cycle method. We consider the following situation: An ideal gas is enclosed in a large vessel under the influence of the gravitational potential $\Phi(z)$. This means that $m\Phi(z)$ is the potential energy of one mole with mass m at the height z above a certain zero level. The gas is in thermal equilibrium with temperature T. The cycle shall be the sequence of the following steps (compare Figure 28). First, one mole is withdrawn by a piston from the vessel at zero level of the potential where pressure be p_0. In the second step, this mole is brought by isothermal expansion from pressure p_0 to the lower pressure p. In the third step it is lifted by mechanical work to the height z where p is the equilibrium pressure in the vessel. Finally the mole of the gas is moved into the vessel by a piston. The cycle, which is now closed, shall be performed reversibly in all steps. The work of the piston involved in withdrawing the mole of the gas from the vessel and in bringing it back to the vessel, both times will have the same amount RT and will cancel out. The work gained by the isothermal expansion is

$$\int_{v_0}^{v} dv\, p = RT \ln \frac{v}{v_0} = RT \ln \frac{p_0}{p} \; . \tag{1.2}$$

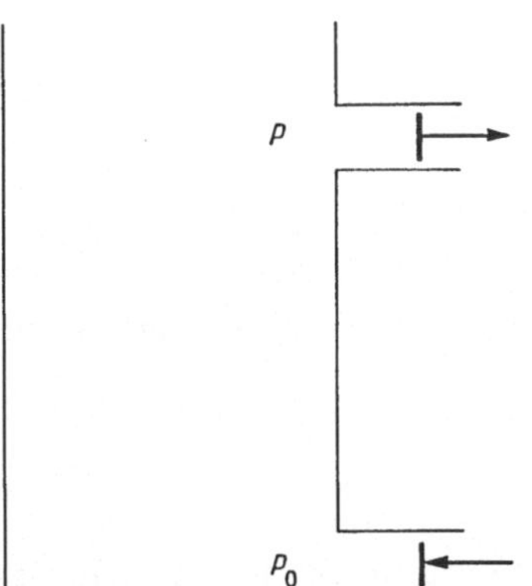

Figure 28

Deduction of the barometric pressure formula with a reversible cycle process. From a vessel in the gravitational field one mole of gas is withdrawn at height z and pressure p. It is shifted back into the vessel at zero height and pressure p_0.

The mechanical work necessary to lift the mole from the zero level to the height z is $m\Phi(z)$. As the total work of the cycle has to be zero, the expression of eq. (1.2) is equal to $m\Phi(z)$. This yields the barometric pressure formula

$$p(z) = p_0 \exp\left(-\frac{m\Phi(z)}{RT}\right). \tag{1.3}$$

3.3.2   Vapor Pressure and Radiation Cavity

The *Clausius-Clapeyron* equation. First we shall discuss the dependence of vapor pressure on temperature. To this purpose we consider the following idealized reversible cycle process. We have two vessels in which liquid and vapor are in thermal equilibrium. One vessel is in contact with a heat reservoir of the temperature T. The other vessel is in contact with another reservoir of the lower temperature $T - dT$. One mole of the vapor with vapor pressure P is reversibly removed from the first vessel by a piston at temperature T. During this process one mole liquid will be evaporized isothermally in the vessel to maintain the equilibrium. The piston then is removed from the vessel. In the second step the gas is expanded adiabatically in the isolated piston from pressure P to pressure $P - dP$ which is the vapor pressure in the second vessel. After this expansion the gas has the lower temperature $T - dT$. In the third step the gas is moved into the second vessel isothermally at temperature $T - dT$ by the piston. This process is connected with a condensation of one mole gas to liquid. In the last step one mole liquid withdrawn from the second vessel is compressed adiabatically from pressure $P - dP$ to the higher pressure P and is finally added to the first vessel. The cycle is closed. The total work gained by the cycle is (compare Figure 29)

$$\oint \mathrm{d}v\, p = (v'' - v')\, \mathrm{d}P. \tag{2.1}$$

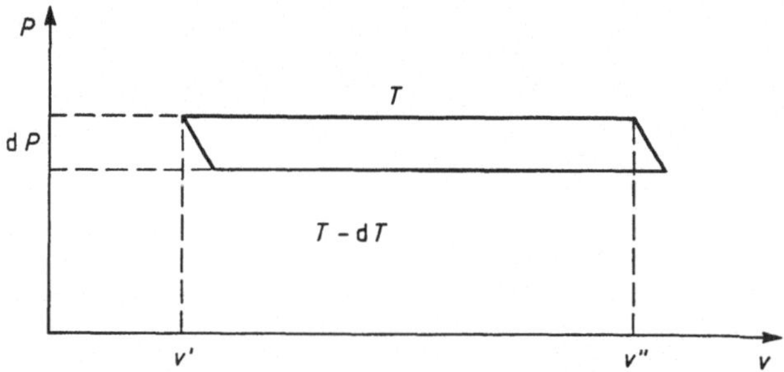

Figure 29 Deduction of the *Clausius-Clapeyron* equation of vapor pressure with a cycle process. The cycle in the (p, v)-diagram.

The difference between gas volume v'' at the two different temperatures is of higher order of dT and is negligible in this equation. The same is true the more for the volume v' of the liquid. The heat withdrawn from the warmer reservoir is the molar evaporation heat q. The maximum efficiency principle thus yields

$$\frac{v'' - v'}{q}\, \mathrm{d}P = \frac{\mathrm{d}T}{T}. \tag{2.2}$$

This is the *Clausius-Clapeyron* equation (7.16) of chapter 2.3.7. The deduction is applicable also to other phase transitions which are associated with a change of the volume of the substance, like melting of a solid. Such transitions are phase transitions of *first order*. The transition heat q is also called *latent heat*.

The *Kirchhoff* law of radiation. In chapter 2.3.5 we considered the equilibrium state of a cavity filled with electromagnetic radiation. We obtained the *Stefan-Boltzmann* law for the spatial energy density u of this radiation. A particular part of this law is that u is a universal function of temperature only, independent of any other property of the cavity. This result was already found by *R. Kirchhoff* in 1859, whereas the special form of the temperature function eq. (5.18) of chapter 2.3.5 was found experimentally by *J. Stefan* in 1879 and deduced theoretically by *L. Boltzmann* in 1884.

First we state that heat capacity eq. (8.24) of chapter 2.3.8 is never negative. Therefore an increase of the energy of the cavity is always connected with an increase of the temperature. Let us imagine that there are two radiation cavities of the same temperature. If the energy density u were different in both cavities, it would be possible to heat up the cavity with the lower value of u and cool the other one by exchange of radiation. Such an exchange could be established by an adequate arrangement of optical lenses to direct the ray emerging from a small hole in one cavity and going in a small hole in the other cavity (compare Figure 30). Energy would flow from the cavity with higher u to that of lower u. The temperature of the first one would decrease, that of the second one would, however, increase. This is in contradiction to the Second Law because it would allow a perpetuo mobile of second kind. Energy density of the radiation in a thermal equilibrium indeed has to depend on temperature only.

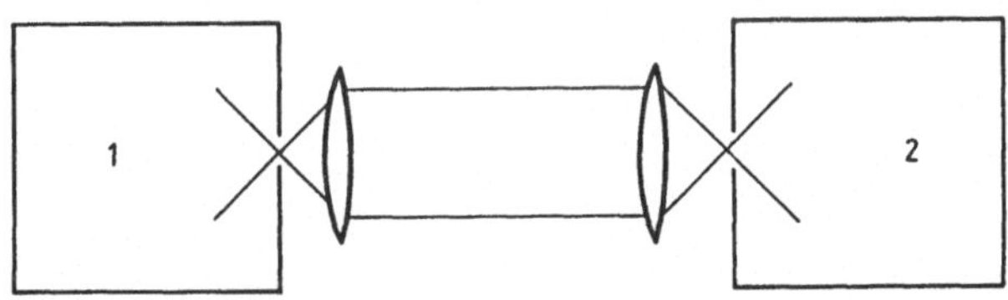

Figure 30 Deduction of the *Kirchhoff* law of radiation. Two radiation cavities in optical exchange of radiation energy.

The *Stefan-Boltzmann* law. Now we can refer to the cycle method to obtain a result about the temperature dependence of u. Let us imagine one wall of the cavity accomplished in the form of a movable piston. In the first step of the cycle we expand the cavity isothermally from volume V' to volume V''. This is done in thermal contact with the heat reservoir of temperature T. Then the piston will be isolated and in the second step adiabatically expanded to the temperature $T - dT$ of another heat reservoir. The third step is compressing the piston, in contact with this reservoir, to a volume from which, as the fourth step, an adiabatic compression leads back to the initial state of the cycle. This cycle is very similar to the cycle used for the derivation of the *Clausius-Clapeyron* equation. We can refer to Figure 29 if we replace the vapor pressure P with the radiation pressure p. As the radiation pressure p is uniquely determined by u, it is also dependent only on T and thus constant on the isothermals. The work gained by the cycle, which in all steps has to be reversible, is again given by eq. (2.1) if p, V, Q is inserted instead of P, v, q. The heat Q withdrawn from the warmer reservoir is determined by the following equation. The work gained by the isothermal expansion at T is

$$(V'' - V')p = -(V'' - V')u + Q. \tag{2.3}$$

Therefore eq. (2.2) assumes the form

$$\frac{\mathrm{d}p}{\mathrm{d}T} = \frac{u + p}{T} \, . \tag{2.4}$$

This equation has already been obtained in chapter 2.3.6, eq. (6.16) and finally yielded the *Stefan-Boltzmann* law

$$u = CT^4 \, . \tag{2.5}$$

The deduction of the *Kichhoff* law can be extended to the proof that the *spectral energy density* $f(T, v)$ is a universal function as well. This quantity is defined by the requirement that the energy density of radiation in the band between the optical frequencies v and $v + \mathrm{d}v$ shall be $f(T, v)\,\mathrm{d}v$. We can insert an optical filter which allows only radiation of this frequency band to pass into the path of the ray between the cavities. If the spectral energy density were different in the cavities, we could heat up one cavity on account of the other one, in contradiction to the Second Law. Therefore, as a consequence of this law, $f(T, v)$ has to be a unique function of T and v.

4 Microscopic Description of Special Systems

In the preceding section 3 we considered applications of macroscopic thermo-dynamic principles to different standard systems. Now we are interested in results which can be obtained only by the application of the statistical theory starting with the individual properties of a system on the microscopic level.

Particular results will be the macroscopic equations of state. In the first subsection we shall be concerned with thermal equations of state. They depend decisively on the forces between the molecules. In the second subsection we shall discuss the specific heat. This quantity comprises just that information over the caloric equation of state which is independent of the thermal one. The caloric properties of a system do not depend only on the intramolecular forces, but considerably more on the number of degrees of freedom. In the third subsection we shall discuss magnetic systems because for these a theory can be based on relatively simple microscopic models, which can be used to explain many characteristic properties of magnets.

4.1 Thermal Equations of State

The macroscopic thermal equation of state of a substance is determined primarily by the intramolecular forces. To describe these forces, we need the micro-scopic level of description. So far as we shall develop the theory in this subsection, it suffices to describe the molecules as mass points.

The first chapter will deal with the ideal gas. This is rarefied in such a degree that the molecules on the average are so far away from one another that the intramole-cular forces have no influence on the thermal equation of state. The molecules can be considered to be uncorrelated. This means that the probability of a microstate of the whole gas is the product of the probabilities of the microstates of the single molecules. We consider identical molecules. Therefore it suffices to study the probability distribu-tion in the phase space of only one molecule.

If we study deviations from the ideal behavior of a gas, we have to include intramolecular interactions and need the probability distributions in the whole high-dimensional Γ-space. It is convenient to deal with the particle number in a fixed partial volume not as a fixed number but as a random quantity. That is, we shall use the grand canonical distribution. This justifies the considerations of the second chapter. In the then subsequent chapters we shall discuss the real gas and a particular model of a liquid.

In states of gases in which quantum mechanics is important, correlations of a type different from those caused by intramolecular forces appear. The new correlations are different for the two possible classes of particles, the *bosons* and the *fermions*. This will be the theme of chapter 5. Chapter 6 is an extension of the theory of a *Bose* gas to the photons of the electromagnetic radiation.

4.1.1   Ideal Gas

The ideal gas is rarefied to an extent that the mean distance of the molecules is large compared with the range of the intramolecular forces. These forces are not totally ineffective. They are responsible for the energy exchange between the molecules by collisions which drive them into a common thermal equilibrium. The time, however, during which a molecule is in the range of the intermolecular forces is small compared to the duration of the free flight. Therefore the mean potential energy of these forces as contribution to the whole energy of the gas can be neglected in comparison with the kinetic energy of the molecules. We assume that no external forces are acting on the molecules, like for instance gravitation. Hence the macroscopic internal energy U of the gas is the mean value of only the kinetic energy of the molecular motion. In the canonical distribution of the gas the *Hamiltonian* appears, but only the kinetic energy has to be respected. The kinetic energy, however, is additive with respect to the single molecules, and the distribution becomes the product of the one-particle distributions of all molecules. We can restrict the discussion to that of a single molecule. Its canonical distribution in the classical Γ-space is

$$\rho = \exp \beta (F_1 - H), \tag{1.1}$$

where H is the classical *Hamilton* function of the single molecule and F_1 is its contribution to the *Helmholtz* free energy of the gas. The *Hamiltonian* is pure kinetic energy. q shall be the whole set of the independent space coordinates of the molecule and P the set of the canonical conjugate momenta. The number of molecules of the gas is N. Then the *Helmholtz* free energy of the gas is

$$F = NF_1 = -\frac{N}{\beta} \ln \int dq \, dP \exp (-\beta H). \tag{1.2}$$

For the free mass point we can for instance choose three *Cartesian* coordinates x_l. Then $\mathbf{H}$ depends on the set P_l only and the integration over $\vec{x}$ yields the volume V as factor in the integral. If we call the remaining part of the integral which depends only on temperature $a(T)$, we obtain

$$F = -NkT[\ln V + \ln a(T)]. \tag{1.3}$$

In this description the volume is a sharp parameter as discussed in chapter 2.2.4. We use eq. (4.8) of 2.2.4 to obtain the pressure

$$p = -\left(\frac{\partial F}{\partial V}\right)_T = \frac{NkT}{V} . \tag{1.4}$$

This is the ideal gas equation. It obtains the familiar form eq. (5.3) of chapter 2.3.5 if we assume N to be the *Loschmidt* number L of molecules in one mole. Eq. (1.4) shows that kL is equal to the gas constant R. This result allows us to interpret the *Boltzmann* constant k as the gas constant per molecule instead of one mole.

The *Maxwell* distribution. With the choice of *Cartesian* coordinates $\vec{x}$, the P_l become the three components of the translational momentum

$$\vec{P} = m\vec{v}, \tag{1.5}$$

where $\vec{v}$ is the velocity of the molecule and m is its mass. If we integrate ρ over $\vec{x}$ only, we obtain the marginal distribution over the velocity

$$w(\vec{v}) = C \exp\left(-\frac{mv^2}{2kT}\right). \tag{1.6}$$

$w(\vec{v})\, d^3v$ is the probability of finding the velocity $\vec{v}$ of a certain molecule in the volume element

$$d^3v = dv_1\, dv_2\, dv_3 \tag{1.7}$$

of the three-dimensional velocity space on point $\vec{v}$. That is to say, it is the probability of finding the three components v_l of $\vec{v}$ in the intervals between v_l and $v_l + dv_l$. The distribution eq. (1.6) is called the *Maxwell distribution*. It is a normal distribution in $\vec{v}$-space. Due to eq. (5.7) of chapter 1.1.5 and in accordance with the equipartition theorem (chapter 2.2.3), we obtain

$$\frac{m}{2}\langle v^2\rangle = \frac{3}{2}\, kT. \tag{1.8}$$

We have to distinguish between the probability distribution of eq. (1.6) and the probability distribution over the absolute value of the velocity. The probability of finding this absolute value in the interval between v and $v + dv$ is obtained by integrating $w(\vec{v})$ over the spherical shell in the $\vec{v}$-space which belongs to this interval of the radius. It is

$$W(v)\, dv = 4\pi v^2 w(\vec{v})\, dv, \tag{1.9}$$

$$W(v) = C' v^2 \exp\left(-\frac{mv^2}{2kT}\right). \tag{1.10}$$

In Figure 31 and Figure 32 the two distributions $w(\vec{v})$ and $W(v)$ are represented. Figure 31 shows the distribution w over the only component v_1 of the vector $\vec{v}$. The maximum of $W(v)$ is reached for the value v_m of v which is associated with the energy

$$\frac{m}{2}\, v_m^2 = kT \tag{1.11}$$

of the molecule. In a large temperature regime this *most probable* value v_m of the velocity is only slightly different from the mean value $\langle v\rangle$ and from the square root of $\langle v^2\rangle$. An increase of temperature T yields a change of $W(v)$ which is more important in the domain of large values v where W is small than it is in the neighborhood of the maximum of W. For instance an increase of 10 % of the absolute temperature involves an increase of the probability of finding the velocity of a molecule six times larger than the average velocity by a factor of some hundred. Therefore special chemical reactions which require large velocities v can become considerably more probable by the temperature increase, whereas the mean energy of the molecules is not very changed, namely by 10 %.

The barometric pressure formula. In this book this formula was already derived in different ways in the macroscopic framework. We shall now obtain it as a result of the canonical distribution of the free molecule. Now the potential energy $m\Phi(\vec{x})$ of the

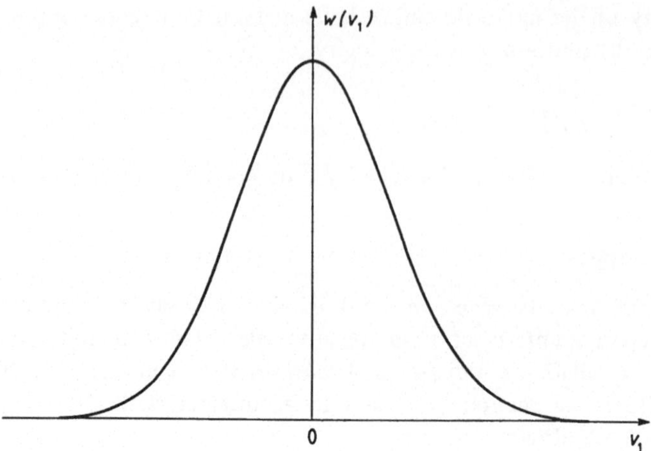

Figure 31 *Maxwell* distribution of one velocity component of the molecules in an ideal gas,

$$w(\vec{v}) = C \exp\left(-\frac{mv^2}{2kT}\right)$$

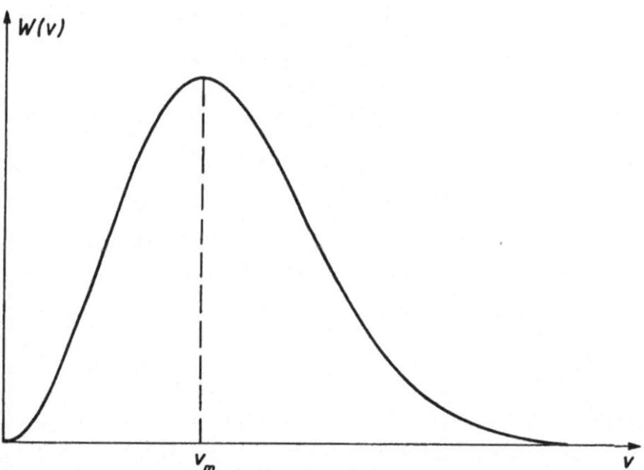

Figure 32 *Maxwell* distribution of the absolute value of the velocity of the molecule in the ideal gas

$$W(v) = C'v^2 \exp\left(-\frac{mv^2}{2kT}\right)$$

molecule in the gravitational field shall be included into the *Hamiltonian H* in the canonical distribution of eq. (1.1). The integration of this equation over only the momenta yields the marginal probability distribution

$$\widetilde{w}(\vec{x})\mathrm{d}^3x = \widetilde{C}\exp\left(-\frac{m\Phi(\vec{x})}{kT}\right)\mathrm{d}^3x \tag{1.12}$$

in $\vec{x}$-space. The particle density of the whole gas is proportional to $\widetilde{w}$. Due to the ideal gas equation this is true for the pressure as well. This result is the barometric pressure formula

$$p(\vec{x}) = p_0 \exp \left(-\frac{m\Phi(\vec{x})}{kT} \right) \tag{1.13}$$

already obtained in the chapters 3.2.4 and 3.3.1.

4.1.2 Grand Canonical Ensemble

The "correct" partition function. The *grand canonical ensemble* has already been introduced in chapter 2.2.3 as one of the special cases of a generalized canonical distribution which we called "standard distributions". It is associated with given mean values of energy **H** and particle numbers **N**. We shall first consider the situation that only one species of particles is present. The distribution is

$$\rho = \exp \beta \left(\Phi - \mathbf{H} + \mu\mathbf{N} \right). \tag{2.1}$$

The constant

$$\Lambda = \exp \left(-\beta\Phi \right) = \operatorname{tr} \exp \beta \left(-\mathbf{H} + \mu\mathbf{N} \right), \tag{2.2}$$

the reciprocal of the normalization constant, is called the *grand partition function*. The name "partition function" has already been used earlier for the corresponding constant

$$\hat{Z}_n = \exp \left(-\beta F_n \right) = \operatorname{tr} \exp \left(-\beta\mathbf{H}_n \right) \tag{2.3}$$

of the canonical distribution. In the canonical distribution the particle number is a sharp parameter and not a random quantity. We are using here the subscript n to indicate that the particle number has this value. The phase space of the grand canonical ensemble is the direct sum of all phase spaces of the n-particle systems with all possible values.

In classical physics, however, the procedure of building up the grand phase space and with this of defining the events with equal "a priori" probability is not unique, as will be explained in the following. If the step from n to $n+1$ were the addition of an individual particle to the system, then

$$\Lambda = \sum_n \zeta^n \hat{Z}_n \tag{2.4}$$

would be the grand canonical partition function, where the quantity

$$\zeta = \exp \left(\beta\mu \right) \tag{2.5}$$

is called *fugacity*. We shall, however, discuss another procedure.

Before we shall do so, let us consider the situation in quantum physics where there are no individual particles. Particles of the same species are in principle indistinguishable. A permutation of the coordinates (including spin variables) of n particles of the same species does not change the microstate. (If the particles are fermions, the wave function can obtain a factor -1, as will be pointed out in chapter 4.1.5. This, however, is not a genuine change of the microstate. In particular, it does not change $\mathbf{H}_n$ and $\hat{Z}_n$.)

This invariance involves a totally different structure of the phase space of the micro-states than that of classical physics. In Λ of eq. (2.4), all permutations of the particle coordinates occur as different microstates, whereas in quantum mechanics they describe one and the same microstate. Therefore in quantum mechanics instead of eq. (2.4), the so-called *"correct" grand partition function*

$$\Lambda = \sum_n \frac{1}{n!}\, \zeta^n \hat{Z}_n \tag{2.6}$$

has to be taken. Correspondingly,

$$Z_n = \frac{1}{n!}\, \hat{Z}_n \tag{2.7}$$

is introduced as the "correct" partition function of the canonical distribution. With this we can write eq. (2.6) in the form

$$\Lambda = \sum_n \zeta^n Z_n. \tag{2.8}$$

In the following we shall show that also in classical physics this is the correct expression, corresponding to realistic situations. The grand canonical ensemble in physics corresponds to a partial system of a larger system of many particles. As a rule it is a part defined by a fixed value of the volume. The particle number in it may be n, whereas the number in the larger system may be K. The addition of an individual particle to establish the step from n to $n + 1$ leading to eq. (2.4) would be unphysical. The correct way is to allow that any subset of n particles may be in the partial volume when performing the canonical partition function $\hat{Z}$. This leads us to

$$\Lambda = \sum_n \binom{K}{n} \zeta^n \hat{Z}_n. \tag{2.9}$$

As n shall run to arbitrarily large values in this sum, we have to assume that K is very large compared to the mean value of n. Thus we may apply the *Stirling* formula

$$K! \approx e^{-K} K^K, \tag{2.10}$$

$$\binom{K}{n} \approx \frac{1}{n!}\, e^{-K} K^n \left(1 - \frac{n}{K}\right)^{n-K}. \tag{2.11}$$

Moreover with

$$\left(1 - \frac{n}{K}\right)^{-K} \approx e^n \tag{2.12}$$

we can put

$$\binom{K}{n} \approx \frac{1}{n!}\, K^n \tag{2.13}$$

in eq. (2.9). This is the same as replacing ζ with $K\zeta$ in eq. (2.8). So we obtain eq. (2.8) also in classical physics.

The generalization to mixtures of different particle species is simple. If n_1, $n_2, \ldots$ are the numbers of particles of species $1, 2, \ldots$ respectively, eq. (2.7) has to be replaced with

$$Z_{n_1 n_2 \ldots} = \frac{1}{n_1! \, n_2! \, \ldots} \hat{Z}_{n_1 n_2 \ldots} \tag{2.14}$$

and eq. (2.8) with

$$\Lambda = \sum_{n_1 n_2 \ldots} \zeta_1^{n_1} \zeta_2^{n_2} \ldots Z_{n_1 n_2 \ldots} \tag{2.15}$$

with

$$\zeta_a = \exp(\beta \mu_a). \tag{2.16}$$

Thermal variables. We return to the pure substance of only one particle species and to eq. (2.9), whichever the basic theory, classical or quantum physics. The mean values of **H** and **N** are the macroscopic thermal variables U and N respectively. We require that the equations

$$TS = -\Phi + U - \mu N, \tag{2.17}$$
$$T\,\mathrm{d}S = \mathrm{d}U - \mu\,\mathrm{d}N \tag{2.18}$$

remain valid. This is the case if we redefine a new μ by replacing ζK with the new fugacity

$$\zeta = \exp(\beta \mu). \tag{2.19}$$

Due to the *Gibbs-Duhem* equation (4.34) of chapter 2.3.4, for the pure substance $N\mu$ is the *Gibbs* free energy G. Therefore we can write eq. (2.17) in the form

$$\Phi = F - \mu N = F - G = -pV, \tag{2.10}$$

in which volume V and pressure p occur.

We have to connect the macroscopic quantities with the grand partition function Λ. The connection with Φ is given by eq. (2.2) with which we obtain the important relation

$$pV = kT \ln \Lambda. \tag{2.21}$$

For the following it is advantageous to choose T, ζ as independent variables in the equation

$$\mathrm{d}\left(\frac{\Phi}{T}\right) = -\frac{U}{T^2}\,\mathrm{d}T - N\,\mathrm{d}\left(\frac{\mu}{T}\right) \tag{2.22}$$

which results from eqs. (2.17, 2.18). With

$$\mathrm{d}\left(\frac{\mu}{T}\right) = k\,\frac{\mathrm{d}\zeta}{\zeta} \tag{2.23}$$

this yields

$$\left(\frac{\partial}{\partial T}\,\frac{\Phi}{T}\right)_{\zeta} = -\frac{U}{T^2}\,,\tag{2.24}$$

$$\left(\frac{\partial}{\partial \zeta}\,\frac{\Phi}{T}\right)_{T} = -k\,\frac{N}{\zeta}\,.\tag{2.25}$$

The first of these two equations can be written in the form

$$U = -\left(\frac{\partial}{\partial\beta}\ln\Lambda\right)_{\zeta},\tag{2.26}$$

the second one in the form

$$N = \zeta\left(\frac{\partial}{\partial\zeta}\ln\Lambda\right)_{\beta}.\tag{2.27}$$

The three relations eqs. (2.21, 2.26, 2.27) will be useful to gain the thermal and caloric equations of state This will be done in the next chapters.

4.1.3   Real Gases

In contrast to the ideal gas, the forces between the molecules in a real gas have an essential influence on the thermal and caloric equation of state, in particular for higher densities. We consider gases for which this is the case at temperatures high enough to ignore quantum effects. Again we describe the molecules as individualizable mass points. Let us enumerate the molecules by a subscript and let r_{lj} be the distance between the l-th and the j-th molecule. For simplicity we assume that the potential energy of the interaction force between them depends only on the distance:

$$\varphi_{lj} = \varphi(r_{lj}).\tag{3.1}$$

If no external forces are present, the classical *Hamilton* function of the gas of molecules is

$$H = \sum_{l=1}^{n}\frac{1}{2m}\,\vec{P}_l^{\,2} + \sum_{l<j}^{n}\varphi_{lj}.\tag{3.2}$$

With the abbreviation

$$\int_{-\infty}^{\infty} d^3 P\,\exp\left(-\frac{\beta}{2m}\,\vec{P}^2\right) = \left(\frac{2\pi}{\beta}\right)^{1/2} = \frac{1}{\lambda}\tag{3.3}$$

we obtain

$$Z_n = \frac{1}{n!}\,\lambda^{-3n}\int d^3x_1 \dots d^3x_n\,\exp\left(-\beta\sum_{l<j}\varphi_{lj}\right).\tag{3.4}$$

In the limit of the sufficiently rarefied gas, we can put φ_{lj} equal to zero. As we shall consider in particular the deviations from an ideal gas in low order, it is convenient to introduce

$$f_{lj} = \exp\left(-\beta\varphi_{lj}\right) - 1, \tag{3.5}$$

which goes to zero in the considered limit. Then

$$Z_n = \frac{1}{n!}\,\lambda^{-3n}\int d^3x_1\ldots d^3x_n \prod_{l<j}^{n}(1+f_{lj}) \tag{3.6}$$

yields the expansion

$$Z_n = \frac{1}{n!}\,\lambda^{-3n}\int d^3x_1\ldots d^3x_n\left(1+\sum f_{lj}+\sum f_{lj}f_{kj}+\ldots\right) \tag{3.7}$$

in powers of f_{lj}. It is also an expansion in powers V^n, V^{n-1}, V^{n-2} ... of the total gas volume V. Such an expansion is called a *virial expansion*. The restriction to the lowest order yields

$$Z_n = \frac{1}{n!}\left(\frac{V}{\lambda^3}\right)^n\left[1+\frac{n(n-1)}{2}\,\frac{1}{V}\int d^3x\, f(r)\right], \tag{3.8}$$

where $\vec{x}$ shall be the oriented distance between two molecules with the length r. Eq. (3.8) is an appropriate approximation for small gas density and higher temperature, with which we can calculate the thermal equation of state by using the general relation

$$\frac{pV}{kT} = \ln\sum_{n=1}^{\infty}\zeta^n Z_n. \tag{3.9}$$

For this purpose let us introduce the following abbreviations:

$$\zeta/\lambda^3 = \xi, \tag{3.10}$$

$$\frac{1}{2}\int d^3x\, f(r) = 2\pi\int_0^{\infty}dr\, r^2 f(r) = q. \tag{3.11}$$

Then we obtain

$$\Lambda = \sum_{n=1}^{\infty}\frac{1}{n!}\,(\xi V)^n\left[1+n(n-1)\frac{q}{V}\right], \tag{3.12}$$

$$\Lambda = e^{\xi V} + \frac{q}{V}\,(\xi V)^2\,\frac{\partial^2}{\partial(\xi V)^2}\,e^{\xi V}, \tag{3.13}$$

$$\Lambda = e^{\xi V}(1+\xi^2 q V). \tag{3.14}$$

According to an approximation for high temperature in which f and thus q is small, we can write

$$\ln \Lambda = \xi V + \xi^2 q V, \tag{3.15}$$

$$\frac{p}{kT} = \xi + q\xi^2. \tag{3.16}$$

Eq. (2.24) for the mean value N of n yields

$$N = \xi \frac{\partial}{\partial \xi} \ln N = \xi V + 2q\xi^2 V. \tag{3.17}$$

In the small term with q, we may approximate ξ by V/N and obtain

$$\xi = \frac{N}{V} - 2q \left(\frac{N}{V}\right)^2, \tag{3.18}$$

$$\frac{p}{kT} = \frac{N}{V} - q \left(\frac{N}{V}\right)^2. \tag{3.19}$$

This is the thermal equation of state. We write it for one mole for which N is the *Loschmidt* number L, and for which we use v instead of V:

$$\frac{pv}{RT} = 1 - \frac{qL}{v}. \tag{3.20}$$

Now we make the following assumption about the form of the interaction potential $\varphi(r)$. It shall be qualitatively of the shape represented in Figure 33. In particular

$$\varphi = \begin{cases} \infty & \text{for} \quad r < d \\ < 0 & \text{for} \quad r > d. \end{cases} \tag{3.21}$$

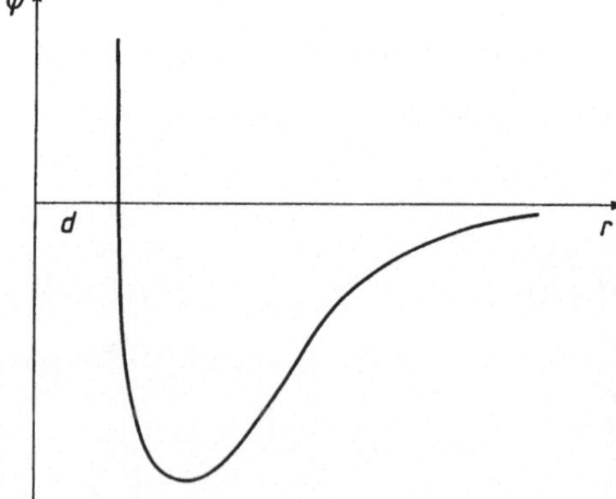

Figure 33

The qualitative type of the intermolecular potential φ as supposed in the discussion of this chapter. r is the distance between two molecules.

In the high temperature approximation $\beta \ll 1$:

$$f = \begin{cases} -1 & \text{for} \quad r < d \\ -\beta \varphi & \text{for} \quad r > d, \end{cases} \tag{3.22}$$

$$q = -\frac{2\pi}{3} d^3 - \frac{2\pi}{kT} \int\limits_d^\infty dr\, r^2 \varphi. \tag{3.23}$$

This results in the thermal state equation

$$pv = RT \left(1 + \frac{b}{v}\right) - \frac{a}{v} \tag{3.24}$$

with:

$$b = L\, \frac{2\pi}{3} d^3, \tag{3.25}$$

$$a = -2\pi L^2 \int\limits_d^\infty dr\, r^2 \varphi. \tag{3.26}$$

The result is in agreement with the *van der Waals* equation

$$p = \frac{RT}{v - b} - \frac{a}{v^2} \tag{3.27}$$

if $b \ll v$, that means for the sufficiently rarefied gas.

The result shows the connection of the parameter a with the attractive part of the interaction and of b with the repulsive part.

*4.1.4 Cell Model of a Liquid

The liquid is a state of matter in which the mean distance between neighboring molecules is not very much larger than in the solid state and is much smaller than in the gaseous state. In the solid state, the molecules are ordered in a crystal lattice over larger distances. The lattice order also is present in some weaker form in the liquid over smaller distances. This order vanishes after several multiples of the lattice constant. An essential property of the liquid is that the molecules have more freedom to move than in the solid crystal.

We shall now be concerned with a relatively simple model of the liquid with in a certain way takes into account this mobility of the molecules. It is the *cell model* [15]. It is able to explain only special features of the liquid. One reason of discussing it here is methodical. Another one is that it is a further example of a microscopic system which in a relatively direct way yields the thermal equation of state.

In this model the motion of a molecule ("wander molecule") is considered in the ordered surroundings of fixed neighboring molecules. We assume individualizable molecules, according to classical mechanics. In contrast to the solid state, the distances between the molecules are a bit larger, thus giving room to the wander molecule in a certain finite domain, a "cell", in which it, however, remains enclosed by the neighbors (Figure 34).

We describe the thermal equilibrium with the canonical distribution of the wander molecule in the cell. This yields an expression for the *Helmholtz* free energy F of the gas in dependence on its volume V. By application of

$$p = -\left(\frac{\partial F}{\partial V}\right)_T \tag{4.1}$$

we obtain the thermal equation of state

$$p = p\,(T,\,V). \tag{4.2}$$

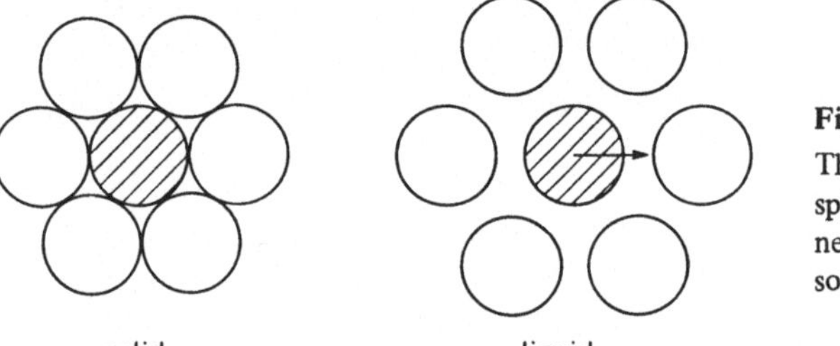

Figure 34

The "wander molecule" (shaded sphere) in the surrounding of neighboring molecules, a) in the solid, b) in the liquid

By use of the *Cartesian* coordinates $\vec{x}$ the classical *Hamilton* function of the wander molecule is separated with respect to position $\vec{x}$ and translational momentum $\vec{P}$:

$$H = \frac{1}{2m}\,\vec{P}^2 + \Phi(\vec{x}) \tag{4.3}$$

The partition function factorizes correspondingly:

$$Z = \int d^3x\, d^3P \exp\,(-\beta H) = Z_P Z_x. \tag{4.4}$$

In this expression only the configuration integral

$$Z_x = \int d^3x \exp\,(-\beta \Phi) \tag{4.5}$$

depends on volume V of the liquid.

The volume of the zone in the cell in which the center of the wander molecule can move shall be ω. Now we make the simplifying assumption that the potential energy Φ of this molecule as function of the position of its center has the constant value ψ in the interior of this zone and is plus infinity outside. Then

$$Z_x = \omega \exp\,(-\beta \psi). \tag{4.6}$$

The term depending on volume V in the free energy F of N molecules is

$$F_V = -NkT \ln Z_x = N(\psi - kT \ln \omega). \tag{4.7}$$

In an approximation we describe the molecule to be a rigid sphere with radius s_0 and the cell to be a sphere with radius s. Then

$$\omega = C(s - s_0)^3 = C\left[\left(\frac{V}{N}\right)^{1/3} - \left(\frac{V_0}{N}\right)^{1/3}\right]^3. \tag{4.8}$$

Thus we obtain

$$F_V = -N\psi - 3NkT \ln(V^{1/3} - V_0^{1/3}) - NkT \ln\frac{C}{3N}, \tag{4.9}$$

$$p = -\left(\frac{\partial F_V}{\partial V}\right)_T = NkT\frac{V^{-2/3}}{V^{1/3} - V_0^{1/3}} + N\frac{\partial \psi}{\partial V}. \tag{4.10}$$

Considering one mole, we put N, V equal to L, v respectively. The value ψ depends on V. It is some kind of spatial mean value of an attractive potential caused by the neighbours. The nearer these are, the stronger the attraction and the lower the negative value ψ. Therefore we make the ansatz

$$L\psi = -\frac{a}{v}. \tag{4.11}$$

We write b instead of v_0 and obtain the *Eyring equation of state*:

$$\left(p + \frac{a}{v^2}\right)(v - b^{1/3}\, v^{2/3}) = RT, \tag{4.12}$$

which resembles the *van der Waals* equation. It proves to be a useful semi-empirical description for liquids if extreme temperatures and other extreme conditions are excluded, in particular if a is allowed to be moderately dependent on temperature.

4.1.5 Perfect *Bose* and *Fermi* Gas

The set of microstates of a system built up of equal particles is decisively different in quantum mechanics from that of a classical gas. A fundamental principle in quantum mechanics is that particles of the same species are not distinguishable. An exchange of all coordinates of two particles including the spin coordinates will not lead to a new microstate of the system. The multiplication of a quantum mechanical wave function with a phase factor $\exp(i\alpha)$ with real α does not change the physical state. Therefore such an exchange of the two particles can change the wave function only by a phase factor. Application of the exchange twice, however, has to leave the wave function unchanged. Thus the only possible phase factors of one exchange of one pair are $+1$ and -1. There are indeed two different classes of particles, the *bosons* with the value $+1$ and the *fermions* with the value -1 of this factor. First it was found empirically and later explained by quantum field theory that, expressed by the natural unit $\hbar$ the bosons always possess integer values of spin and the fermions odd multiples of $1/2$ (this unit $\hbar$ is *Planck*'s constant h divided by 2π). The behavior of the first class is called *Bose-Einstein statistics*, that of the latter *Fermi-Dirac statistics*.

In the following we shall consider *perfect gases*. These are gases of particles with no mutual interaction forces. In classical physics we call them "ideal gases". In quantum mechanics we prefer the term "perfect" because there the usual "ideal" gas equation is no longer valid. A complete orthogonal basis of quantum mechanical states in *Hilbert* space of such a gas can be described by the set of all particle numbers n_i in the one-particle states $|i\rangle$. These one-particle states form a complete orthogonal basis of the one-particle *Hilbert* space. The distinction between bosons and fermions now is the following: for bosons any integer positive number including zero is allowed to be n_i; for fermions, however, only the values 0 and 1 are possible. It should be stressed that the one-particle state $|i\rangle$ includes the spin state of the particle. For the perfect gas, a many-particle state with well defined quantum numbers n_i is an eigenstate of energy of the whole gas. For an imperfect gas, these many-particle states also form a complete basis of the many-particle *Hilbert* space but in general they are not stationary states with a sharp value of total energy. They become nearly stationary quantum states, however, if the intramolecular forces are weak. Then the use of this basis is adequate if the interaction can be handled as a perturbation.

The selection of the microstates which belong to a fixed total number of particles, and simultaneously satisfy the restrictive conditions for the numbers n_i, is complicated. Therefore we shall describe the thermal equilibrium state of a fixed partial volume V with a grand canonical distribution. Particle number **N** and *Hamiltonian* **H** are commutable variables. The statistical operator ρ thus has the same eigenstates like these observables. Therefore we can choose as the one-particles states $|i\rangle$ the energy eigenstates of a particle with energy ϵ_i. The grand canonical partition function eq. (2.2) then is

$$\Lambda = \prod_i \Lambda_i \tag{5.1}$$

with

$$\Lambda_i = \sum_n \exp\left[\beta(\mu - \epsilon_i)n\right]. \tag{5.2}$$

It is substantially different for bosons, for which n runs through all positive integers including zero, from that for fermions, for which n assumes only the values 0 and 1. So we obtain for bosons:

$$\Lambda_i = \{1 - \exp\left[\beta(\mu - \epsilon_i)\right]\}^{-1} \tag{5.3}$$

and for fermions:

$$\Lambda_i = 1 + \exp\left[\beta(\mu - \epsilon_i)\right]. \tag{5.4}$$

As Λ_i has to be finite, μ is restricted for bosons by

$$\mu < \epsilon_i, \tag{5.5}$$

whereas it is unrestricted for fermions.

We shall write uniformly

$$\Lambda_i = \{1 \mp \exp\left[\beta(\mu - \epsilon_i)\right]\}^{\mp 1} \tag{5.6}$$

and set down here and for the following that the upper sign holds for bosons, the *Bose gas*, and the lower sign for fermions, the *Fermi gas*. The mean value of n_i is

$$N_i = \frac{1}{\beta}\,\frac{\partial}{\partial\mu}\,\ln\Lambda_i = \{\exp[\beta(\epsilon_i - \mu)] \mp 1\}^{-1}. \tag{5.7}$$

This quantity is plotted as a function of ϵ_i in Figure 35 for the *Bose* gas, and in Figure 36 for the *Fermi* gas at two different temperatures. The expression for the *Fermi* gas becomes a sharp step function in the limit of absolute temperature zero. In this limit all

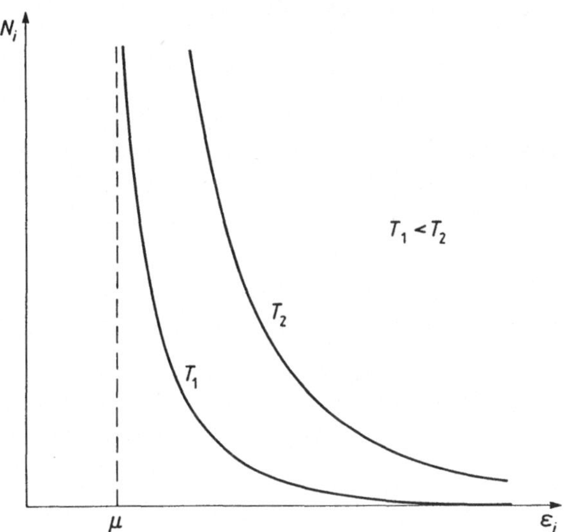

Figure 35 Particle number N_i in a one-particle state with energy ϵ_i for the perfect *Bose* gas

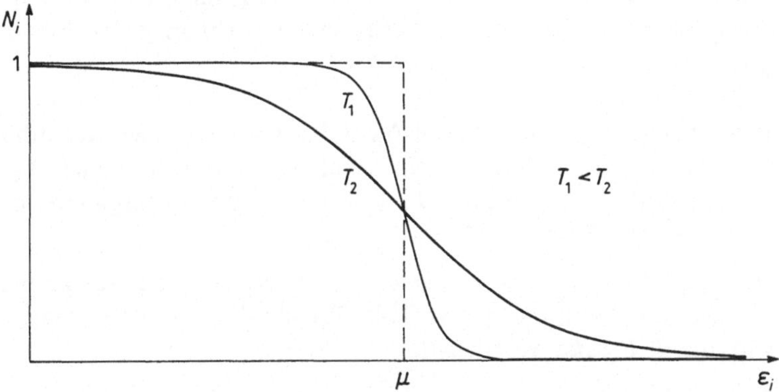

Figure 36 Particle number N_i in a one-particle state with energy ϵ_i for the perfect *Fermi* gas

one-particle states with $\epsilon_i > \mu$ are occupied with one particle, all others are empty. The total particle number is on the average

$$N = \sum_i N_i \,, \tag{5.8}$$

and due to eq. (2.21), we obtain the relation

$$\frac{p}{kT} = \frac{1}{V} \sum_i \ln \Lambda_i \,. \tag{5.9}$$

If the gas is highly rarefied, that is to say, if $N_i \ll 1$, the restrictions for the occupation numbers become unimportant and the classical value

$$N_i = \exp\left[\beta\left(\mu - \epsilon_i\right)\right] \tag{5.10}$$

will be approached.

Ground level of the *Bose* gas. The *Bose* gas requires a special consideration with respect to the lowest one-particle level ϵ_0. For convenience we can define this level to be zero by an adequate definition of μ. With introduction of the fugacity

$$\zeta = \exp\left(\beta\mu\right) \tag{5.11}$$

we write

$$N_i = \left[\zeta^{-1} \exp\left(\beta\epsilon_i\right) - 1\right]^{-1} \tag{5.12}$$

and in particular

$$N_0 = \frac{\zeta}{1 - \zeta} \,. \tag{5.13}$$

This number N_0 can become "macroscopic", that means of the order of the mean value of all N_i in a macroscopic system. N_0 is say of order 10^{20} if $1 - \zeta$ is of order 10^{-20}. Then for all $\epsilon_i > 0$ practically

$$N_i = \left[\exp\left(\beta\epsilon_i\right) - 1\right]^{-1} \,. \tag{5.14}$$

This means that if already the first level $\epsilon_1 > 0$ over the ground level is occupied, then all N_i with exception of N_0 become practically independent of ζ. We can thus write

$$N = N' + N_0 \,, \tag{5.15}$$

where N' is independent of ζ and hence of N_0 if N_0 is macroscopic. According to this situation, we may discuss separately the regime of temperature in which N_0 vanishes on the macroscopic scale. In this regime, $1 - \zeta$ is finite and the energy level ϵ_0 may be neglected. Then we say, the *Bose* gas is in the *normal phase*.

The free particle states. We characterize the state $|i\rangle$ of a free particle by the translational momentum $\vec{P}$ and the spin variable. The absolute value P of the momentum is associated with the *de Broglie* wave length

$$\lambda = \frac{h}{P} \,, \tag{5.16}$$

where h is *Planck*'s constant. To perform the summation over all states $|i\rangle$, we have to count the eigenvalues P, that is to say, the standing waves in a given macroscopic volume V. We can assume this volume to be that of a cube with the side length l. Let us characterize the direction of $\vec{P}$ by a unit vector $\vec{\alpha}$ in

$$\vec{P} = P\vec{\alpha}. \tag{5.17}$$

The three components α_j of $\vec{\alpha}$ have to satisfy

$$\frac{\lambda}{2\alpha_j} = \frac{l}{n_j} \tag{5.18}$$

with integer numbers n_j for the standing waves. These are thus characterized by triplets of positive integer numbers

$$\{n_1, n_2, n_3\} = \frac{2l}{\lambda}\vec{\alpha} = \frac{2l}{h}\vec{P} \tag{5.19}$$

which form a simple cubic lattice in a three-dimensional mapping space $\vec{n}$. The number of lattice points corresponding to eigenvalues of $\vec{P}$ in a volume element d^3P of momentum space is

$$\mathrm{d}z = \mathrm{d}^3 n = \left(\frac{2}{h}\right)^3 V\,\mathrm{d}^3 P. \tag{5.20}$$

We suppose that $\mathrm{d}^3 n$ already includes many lattice points and that therefore it is practically equal to their number. In the last equation we have already replaced l^3 with the volume V. As all n_j are positive, and lying in a octant of $\vec{n}$-space, we have to integrate over the corresponding octant in the $\vec{P}$-space. In the following we may extend the integration over the whole $\vec{P}$-space if we divide the result by 8. If σ is the number of possible spin states of one particle, we finally have to put

$$\sum_i \ldots = \sigma V h^{-3} \int \mathrm{d}^3 P \ldots. \tag{5.21}$$

Eq. (5.9) then takes the form

$$\frac{p}{kT} = \sigma\,\frac{V}{h^3} \int \mathrm{d}^3 P \ln \Lambda(P) \tag{5.22}$$

with

$$\Lambda(P) = \{1 \mp \zeta \exp[-\beta\epsilon(P)]\}^{\mp 1}. \tag{5.23}$$

In the case of the *Bose* gas we suppose the restriction to the "normal phase". With the mole volume v we write

$$\frac{L}{v} = \frac{N}{V} = \frac{1}{V}\sum_i N_i \tag{5.24}$$

and obtain

$$\frac{L}{v} = \sigma \frac{4\pi}{h^3} \int_0^\infty dP\, P^2 N(P), \tag{5.25}$$

$$\frac{p}{kT} = \sigma \frac{4\pi}{h^3} \int_0^\infty dP\, P^2 \ln \Lambda(P), \tag{5.26}$$

where $\Lambda(P)$ is given by eq. (5.23) and where

$$N(P) = \frac{1}{\zeta^{-1} \exp\left[\beta\epsilon(P)\right] \mp 1}. \tag{5.27}$$

Nonrelativistic rarefied gas. We shall now consider the gas in which the particle velocity is always small compared to the velocity of light. Then

$$\epsilon(P) = \frac{P^2}{2m}. \tag{5.28}$$

The quantity

$$\lambda = \frac{h}{(2\pi m kT)^{1/2}} \tag{5.29}$$

is by convention defined as *thermal wave length*. With this we write

$$\frac{L}{v} = \frac{4}{\pi^{1/2}} \frac{\sigma}{\lambda^3} \int_0^\infty dx\, \frac{x^2}{\zeta^{-1} \exp x^2 \mp 1}, \tag{5.30}$$

$$\frac{p}{kT} = \frac{4}{\pi^{1/2}} \frac{\sigma}{\lambda^3} \int_0^\infty dx\, x^2 \ln\left[1 \mp \zeta \exp\left(-x^2\right)\right], \tag{5.31}$$

where x is $\beta\epsilon$.

These equations can be evaluated with an expansion in powers of ζ. We shall consider only highly rarefied gases with values $N_i \ll 1$. Due to eqs. (5.7, 511) then $\zeta \ll 1$ and we can break up the expansion with the first power:

$$\frac{L}{v} = \frac{\sigma\zeta}{\lambda^3}, \tag{5.32}$$

$$\frac{pv}{RT} = 1 \mp 2^{-5/2}\zeta. \tag{5.33}$$

This yields a thermal equation of state

$$\frac{pv}{RT} = 1 \mp 2^{-5/2} \frac{\lambda^3}{\sigma v}. \tag{5.34}$$

In comparison to the classical ideal gas, we obtain a *Bose attraction* and a *Fermi repulsion*.

The *Einstein* condensation. The last results for the *Bose* gas are valid in the temperature regime of the "normal phase" in which N_0 remains negligible. If, however, N_0 contributes to

$$N = L \frac{V}{v} = N' + N_0 \tag{5.35}$$

on the macroscopic scale, then we can put ζ equal to 1 in N', which thus becomes independent of ζ, and due to eq. (5.32) we obtain

$$\frac{N'}{V} = \frac{\sigma}{\lambda^3} . \tag{5.36}$$

With eq. (5.29) we see that N'/V is proportional to $T^{3/2}$. This leads us to

$$\frac{N_0}{N} = \begin{cases} 1 - (T/T_c)^{3/2} & \text{for } T < T_c \\ 0 & \text{for } T > T_c, \end{cases} \tag{5.37}$$

where T_c is given by $\lambda^3 = \sigma$ (see Figure 37).

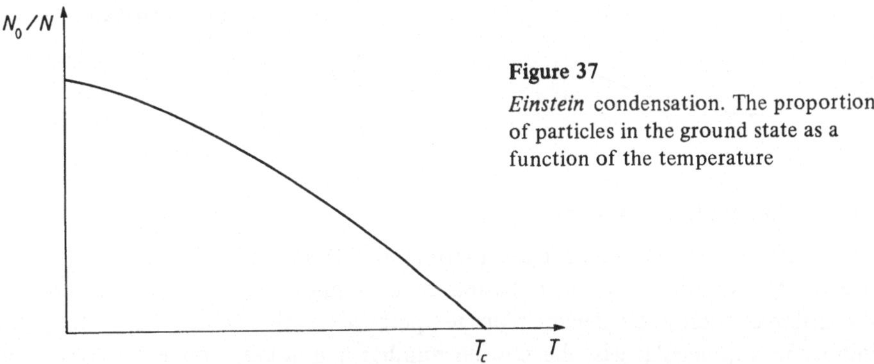

Figure 37

Einstein condensation. The proportion of particles in the ground state as a function of the temperature

If N_0 becomes macroscopic, the contribution of the ground level ϵ_0 to eq. (5.9) is practically

$$-\frac{1}{V} \ln (1 - \zeta) = \frac{1}{V} \ln N_0 \tag{5.38}$$

and thus negligible compared with the rest, which is of the order of L/v as eqs. (5.30, 5.31) show. This means, however, that this rest is of the order of N_0/V. Then with ζ equal to one, eq. (5.31) yields

$$\frac{p}{kT} = \frac{\gamma}{\lambda^3} \tag{5.39}$$

with constant γ. Thus pressure p becomes proportional to $T^{5/2}$. The main result is that pressure is then dependent on temperature only, not on volume, like vapor pressure over a liquid (see Figure 38). Below a temperature T_c the whole many particle system is a composition of two contributions. These two parts are not separated in space, like two components of a mixture. One is the *normal part* of a gas with density N'/V independent of volume, the other is the *condensed part* of N_0 particles. The pressure is only determined by the normal part. The transition from the normal phase to the state in which the normal part and the condensed part are present occurs at the critical temperature T_c. The occurrence of the condensed part is called the *Einstein condensation*. It is important for the interpretation of the superfluidity of helium (precisely of ^{4}He) at low temperatures. The superfluid phase of helium has a certain resemblance to a mixture of two fluids, represented by the normal and the condensed part, respectively.

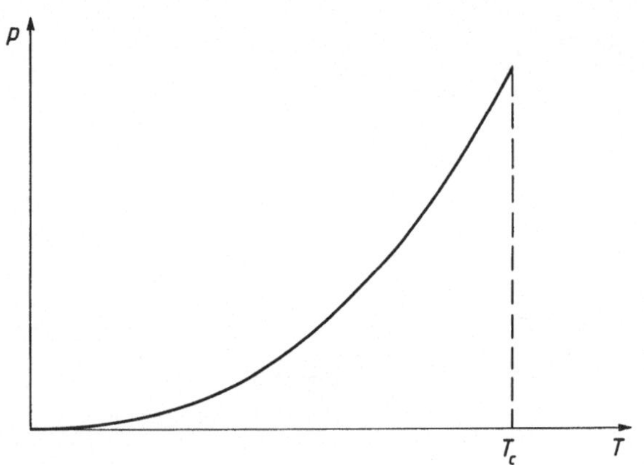

Figure 38

Pressure p of the perfect *Bose* gas in the temperature regime of the *Einstein* condensation

4.1.6   The Radiation Cavity

We shall now consider the thermal equilibrium of the electromagnetic radiation in a cavity by describing the radiation to be a photon gas. The mean photon number, like internal energy U, depends on temperature only. Therefore, unlike the particle number in a molecular gas, the photon number is uniquely connected with U and is not an independent extensity. In this case we have to choose the canonical distribution instead of the grand canonical one. We obtain it by putting μ equal to zero in the expressions for the grand canonical distribution of the *Bose* gas in the preceding chapter. This means that we have to put ζ equal to 1. Photons are bosons with spin 1. In accordance with the existence of only two polarization states of electromagnetic waves, there are, however, only two spin states of the photon. A third spin state is excluded, in accordance with the fact that there are no "longitudinal photons". Moreover, photons have zero mass. Therefore

$$\sigma = 2, \tag{6.1}$$

$$P = \frac{h\nu}{c}, \tag{6.2}$$

$$\epsilon = h\nu, \tag{6.3}$$

where h ist *Planck*'s constant and ν is the optical frequency associated with the wave length

$$\lambda = \frac{c}{\nu} = \frac{h}{P} .$$
(6.4)

The number of standing waves with a frequency in the interval between the values ν and $\nu + d\nu$ thus is

$$dz = 8\pi\nu^2 c^{-3}\, d\nu.$$
(6.5)

Eq. (5.21) now reads

$$\sum_i{}' \ldots = 8\pi c^{-3} \int_0^\infty d\nu\,\nu^2 \ldots .$$
(6.6)

Planck's radiation formula. The spatial energy density of the radiation is

$$u = V^{-1} \sum_i{}' N_i \epsilon_i .$$
(6.7)

In this expression we have to insert the expression of eq. (5.27) with vanishing μ for N_i and obtain

$$N(P) = [\exp(\beta h\nu) - 1]^{-1} .$$
(6.8)

This yields

$$u = \int_0^\infty d\nu\, f(T, \nu)$$
(6.9)

with *Planck*'s radiation formula

$$f(T, \nu) = \frac{8\pi}{c^3}\, \nu^2\, \frac{h\nu}{\exp(h\nu/kT) - 1} .$$
(6.10)

$f(T,\nu)\, d\nu$ is the energy per unit of volume in the frequency band between ν and $\nu + d\nu$.

Radiation pressure. The grand partition function Λ in our case reduces to the canonical partition function and with it

$$F = -kT \ln \Lambda$$
(6.11)

to the *Helmholtz* free energy. If we put μ in eq. (5.6) equal to zero and take the sign for bosons, we are led to

$$F = kT \sum_i \ln[1 - \exp(-\beta\epsilon_i)].$$
(6.12)

The radiation pressure is obtained by

$$p = -\left(\frac{\partial F}{\partial V}\right)_T .$$
(6.13)

If the volume of the cavity is changed, the wave lengths of the standing waves will change. If we enlarge the linear dimensions of the cavity by a scaling factor, all these wave lengths will be enlarged by the same factor. Therefore the eigenfrequencies ν_i will be proportional to $V^{-1/3}$. The same is true for the energy values of the photons

$$\epsilon_i = h\nu_i. \tag{6.14}$$

Thus

$$\frac{\partial \epsilon_i}{\partial V} = -\frac{1}{3}\frac{\epsilon_i}{V}. \tag{6.15}$$

Eqs. (6.12, 6.13) therefore yield

$$p = \frac{1}{3V} \sum_i \frac{\epsilon_i \exp(-\beta\epsilon_i)}{1 - \exp(-\beta\epsilon_i)}. \tag{6.16}$$

With

$$N_i = [\exp(\beta h\nu_i) - 1]^{-1} \tag{6.17}$$

and with eq. (6.7) this is

$$p = \frac{1}{3V} \sum_i N_i \epsilon_i = \frac{u}{3}. \tag{6.18}$$

In this manner we deduced now the relation between radiation pressure p and energy density u of the electromagnetic radiation in thermal equilibrium of eq. (5.17) of chapter 2.3.5 where it was given without proof.

4.2   Specific Heat

In this subsection we shall consider the specific heat of pure substances. We shall restrict the discussion to c_V, the specific heat at constant volume. As we have already pointed out, this specific heat gives an information about the caloric equation of state and is independent of the thermal equation of state. Accordingly, to obtain it experimentally, so-called "caloric" measurements are necessary which are independent from the measurements yielding the thermal equation.

Our aim will be to explain the value of specific heat and its dependence on the basis of the individual microscopic structure of the considered substances. Due to eq. (9.5) of chapter 2.3.9, specific heat at fixed volume is the derivative of the internal energy u per mol with respect to temperature,

$$c_V = \left(\frac{\partial u}{\partial T}\right)_V. \tag{1}$$

It is positive on account of the stability conditions of thermal equilibrium. The positivity can be seen still more directly by the fact that (up to the choice of units) c_V is also the bit number variance of the canonical distribution, as explained in the same chapter 2.3.9.

In the first chapter of this subsection we shall be concerned with the specific heat of ideal gases. It is the sum of two different contributions. The first one arises from the kinetic energy of the translational and rotational motion of the molecules. The second one arises from the internal vibrations of the molecules if they contain more than one atom. These vibrations are usually not observed in the "normal" regime of temperature because of quantum effects. We shall discuss this phenomenon in detail for molecules with only two atoms. Quantum effects also occur with respect to the rotational motion. They, however, become observable at low temperature only. This will be the subject of the second chapter. In the third chapter we shall consider the specific heat of solids. Here quantum mechanics will prove important as well.

4.2.1 Specific Heat of Ideal Gases

The normal regime. As already pointed out, the specific heat of an ideal gas is constant in a large regime of temperature, the so-called "normal regime". Deviations occur only at temperatures which are high or low compared with room temperature. The behavior in the normal regime is what we expect in the framework of classical mechanics of rigid molecules. The deviations can be explained by quantum mechanics of the internal motion of the molecules.

We start with the "normal regime" in which the description of the single molecule as a rigid body proves successful. In the ideal gas the molecules can be assumed to be uncorrelated with one another. This was already discussed in chapter 4.1.1 in connection with the thermal equation of state. Therefore, all thermodynamic extensities of the gas are the sum of the one-particle values. In particular this is true for the heat capacity. In chapter 2.2.3 we obtained the equipartition theorem as a consequence of classical mechanics. It reads that each degree of freedom contributes with the same amount $kT/2$ to the mean value of kinetic energy in a thermal equilibrium. The potential energy of intramolecular forces is negligible. Thus for a mole of L molecules each of which possesses f degrees of freedom, the theorem yields

$$u = fL\frac{kT}{2} = f\frac{RT}{2}, \tag{1.1}$$

$$c_V = f\frac{R}{2}. \tag{1.2}$$

f is 3 for a one-atomic gas, like for a noble gas; it is 5 for a two-atomic gas, like H_2 or HCl; it is 6 for molecules with three or more atoms, like H_2O, NH_3 and so on. The empirical verification of eq. (1.2) gives the two confirmations that in the normal regime the motion of the molecule can be described by classical mechanics and that in this regime the internal motions do not take part in c_V.

The freezing of vibrations. The internal vibrations of the atomic nuclei in the molecule become important at high temperatures. To study them, we shall consider the simplest case of a two-atomic gas. The potential energy $\varphi(r)$ of the interaction between the two nuclei as a function of their distance r is qualitatively expected to be of the shape represented by the solid line in Figure 39. There D is the *dissociation energy*, the difference between the energy zero at which the two atoms become free,

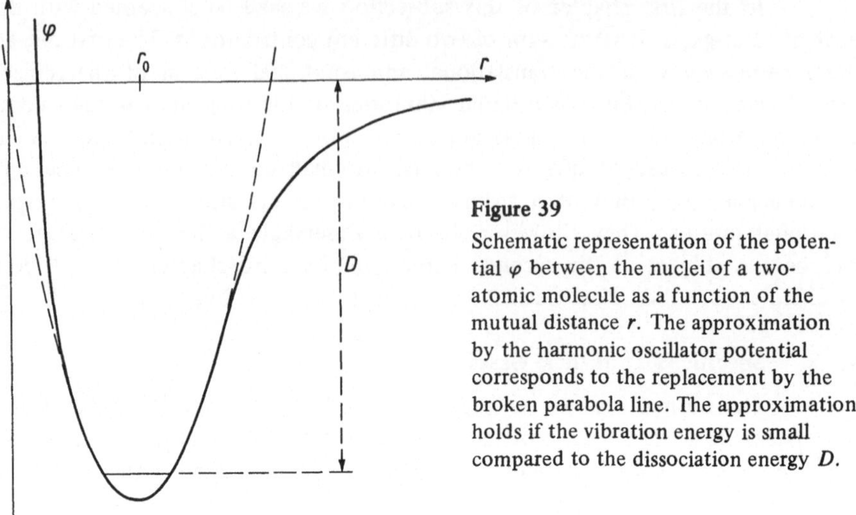

Figure 39
Schematic representation of the potential φ between the nuclei of a two-atomic molecule as a function of the mutual distance r. The approximation by the harmonic oscillator potential corresponds to the replacement by the broken parabola line. The approximation holds if the vibration energy is small compared to the dissociation energy D.

and the lowest bound level. We assume that at the considered temperatures, where we observe the specific heat, the molecules do not practically dissociate. This means that the dissociation is very unprobable and the mean vibrational energy is small compared to D. Then we are allowed to approximate the potential by the quadratic potential

$$\varphi(r) = \frac{c}{2}\,(r - r_0)^2 \tag{1.3}$$

of the linear oscillator, represented by the broken parabola line in Figure 39. With the reduced mass m of the two nuclei, the oscillator frequency is

$$\nu = \frac{1}{2\pi}\left(\frac{c}{m}\right)^{1/2}. \tag{1.4}$$

(The electron masses are neglected as small.) The discrete energy levels of the oscillator are

$$\epsilon_n = \left(n + \frac{1}{2}\right) h\nu \tag{1.5}$$

with positive integer n including zero. As the vibrational motion is independent of the translational motion of the molecule, we may calculate separately the mean energy of the oscillator as an additive contribution to the total energy of the molecule. The canonical partition function of the oscillator is

$$Z = \sum_{n=0}^{\infty} \exp\left(-\beta\epsilon_n\right) = \exp\left(-\alpha/2\right) \sum_{n=0}^{\infty} \exp\left(-n\alpha\right) \tag{1.6}$$

with

$$\alpha = \beta h\nu = \frac{\Theta}{T}, \qquad \Theta = \frac{h\nu}{k}. \tag{1.7}$$

That is

$$Z = \frac{e^{\alpha/2}}{e^{\alpha} - 1}. \tag{1.8}$$

The mean vibrational energy of one molecule thus is

$$\langle \epsilon \rangle = -\frac{1}{Z}\frac{\partial Z}{\partial \beta} = -\frac{h\nu}{Z}\frac{\partial Z}{\partial \alpha}, \tag{1.9}$$

$$\langle \epsilon \rangle = \left(\frac{1}{e^{\alpha} - 1} + \frac{1}{2}\right) h\nu. \tag{1.10}$$

The additive contribution of the vibration to the molar specific heat c_V is called the *vibrational heat*

$$c_{vib} = L\frac{\partial}{\partial T}\langle \epsilon \rangle. \tag{1.11}$$

With

$$\frac{\partial \alpha}{\partial T} = -\frac{\alpha}{T} \tag{1.12}$$

we obtain

$$c_{vib} = -R\alpha^2 \frac{\partial}{\partial \alpha}\frac{1}{e^{\alpha} - 1}, \tag{1.13}$$

$$c_{vib} = R\left(\frac{\Theta}{T}\right)^2 \frac{\exp \Theta/T}{(\exp \Theta/T - 1)^2}. \tag{1.14}$$

This result is represented in Figure 40. The value of c_{vib} for T equal to Θ is $0.92\,R$. For $T \gg \Theta$ it assumes the classical value R. It vanishes like $\exp(-\Theta/T)$ for $T \to 0$.

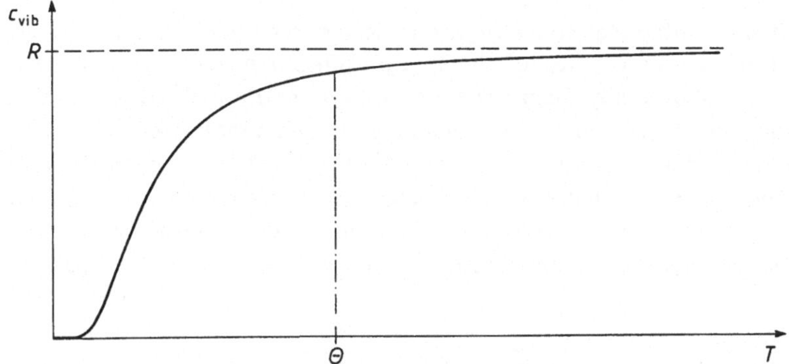

Figure 40 Vibrational heat c_{vib} of a diatomic gas as a function of temperature T. The quantum mechanical "freezing" of the vibrational degree of freedom begins with cooling below the characteristic temperature Θ.

The characteristic temperature Θ is highest for the light molecule hydrogen H_2, in fact about 6100 °K. It is extraordinarily low for the heavy molecule iodine I_2, about 305 °K. The theoretical result eq. (1.14) is in good agreement with observations, and the values of Θ fit the values of the frequencies ν in the case that these are observable optically.

For molecules with more than two atoms, each normal vibration contributes to the specific heat with a term of eq. (1.14) and a corresponding Θ. The typical quantum effect that the vibrations vanish for low temperature is often called the *freezing* of the vibrational degrees of freedom.

4.2.2 Ortho- and Para-Hydrogen

Whereas the quantum character of the internal vibrations of the molecule causes deviations from the classical equipartition theorem at high temperature, the corresponding is true for the rotation of the molecule at low temperature. The influence of rotational quantum effects can best be observed in the lightest molecule with more than one atom, the hydrogen molecule H_2. These observations were very important for the development of general physics because they were associated with the discovery of the nuclear spin and of the fact that protons obey the *Fermi* statistics.

At low energies, where the quantum effects become relevant for the rotation of the molecule, the vibrational degrees of freedom are already totally "frozen" and the molecule rotates as a rigid body. That is to say, it remains in the ground state with respect to vibrations and to electron orbits. The moment of inertia of the H_2 molecule round an axis perpendicular to the connection between the two protons shall be I. Then the levels of the rotational energy are

$$\epsilon_l = \frac{\hbar^2}{2I}\, l(l+1) \tag{2.1}$$

with positive integer l including zero. Each quantum number l is associated with the multiplicity

$$g_l = \sigma_l(2l+1) \tag{2.2}$$

of different pure rotational states of the molecule, including nuclear spin states. σ_l is the number of the spin states of the two protons. It depends on l due to the *Pauli* principle, which allows only quantum states antisymmetric with respect to the exchange of all coordinates of the two protons including the spin coordinates, because the protons are fermions. The states with even l are symmetric in the space coordinates of the nuclei, those with odd l are antisymmetric. This yields that the even l are associated with nuclear spin 1 and σ_l equal 3, and that the odd l are associated with nuclear spin 0 and σ_l equal 1.

The canonical partition function of nuclear spin and of the rotational degrees of freedom is thus

$$Z = \sum_{l=0}^{\infty} g_l \exp\left(-\beta\epsilon_l\right) = \sum_{l=0}^{\infty} \sigma_l(2l+1)\exp\left[-\alpha l(l+1)\right], \tag{2.3}$$

where

$$\alpha = \hbar^2 \beta / 2I = \Theta / T, \tag{2.4}$$

$$\Theta = \hbar^2 / 2Ik. \tag{2.5}$$

We can approximately regard the H_2 gas as a mixture of two different species, the *ortho-hydrogen* with molecules in rotational states with even l and the *para-hydrogen* with odd l because the conversion from one to the other state is a rare event. The partition function Z can be written as the sum of the two partition functions of these two "species",

$$Z_{\text{or}} = 3 \sum_{\text{even } l} (2l + 1) \exp\left[-\alpha l(l + 1)\right] \tag{2.6}$$

of the ortho-states and

$$Z_{\text{par}} = \sum_{\text{odd } l} (2l + 1) \exp\left[-\alpha l(l + 1)\right] \tag{2.7}$$

of the para-states. $Z_{\text{or}} : Z_{\text{par}}$ is the ratio of the two species in the mixture.

At high temperature $T \gg \Theta$ the large l contribute essentially and the ratio $Z_{\text{or}} : Z_{\text{par}}$ is $3:1$. At low temperature $T \ll \Theta$ only the first term with l equal to zero contributes, and Z_{par} is large compared to Z_{or}. Then para-hydrogen is present nearly pure in equilibrium. The conversion of para- to ortho-hydrogen occurs so slowly that by faster heating up the para-hydrogen remains in an inhibited thermal state. Therefore the specific heat of the para-hydrogen can be measured separately if one starts at a very low temperature.

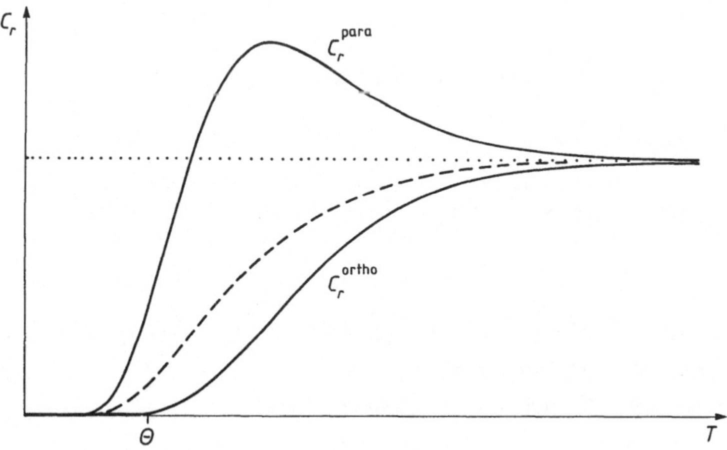

Figure 41 Rotational heat c_r of the hydrogene molecule at low temperature for ortho-hydrogene and para-hydrogene as a function of temperature T, represented as solid lines. The values for the equilibrium mixture are represented as the broken line.

The contribution of the rotational degrees of freedom to the specific heat c_V is called *rotational heat*. It is

$$c_r = L \frac{\partial}{\partial T} \langle \epsilon \rangle \tag{2.8}$$

with

$$\langle \epsilon \rangle = -\frac{1}{Z} \frac{\partial Z}{\partial \beta} . \tag{2.9}$$

In Figure 41 the theoretical values of c_r are plotted as a function of T. They are plotted separately for ortho- and para-hydrogen by solid lines. The values for the mixture in the unrestricted equilibrium are represented by the broken line.

As already mentioned, the different behavior of orth- and para-hydrogen was the first indication that protons possess a spin 1/2 and obey the *Pauli* principle. It was the first indication of the existence of a nuclear spin.

4.2.3   Specific Heat of Solids

The *Dulong-Petit* rule. Also for the solid matter we can define a *normal regime* of temperature in which specific heat can be explained by classical mechanics. Each molecule of the solid shall be described to be a mass point with three degrees of freedom which is bound to the other molecules by harmonic forces. Then one mole of the solid matter is represented in dynamics by $3L$ linear harmonic oscillators, corresponding to the $3L$ normal vibrations. These oscillators are not correlated with one another. Therefore we may first consider the canonical distribution of a single one-dimensional oscillator. Its mean kinetic energy is $kT/2$ due to the equipartition theorem. Its potential energy is

$$\varphi(x) = \frac{\kappa}{2} x^2 . \tag{3.1}$$

The probability density of the one-dimensional coordinate x is

$$\rho(x) = A \exp\left(-\beta \frac{\kappa}{2} x^2\right). \tag{3.2}$$

It is a normal distribution and due to eq. (5.7) of chapter 1.1.5, we obtain

$$\langle \varphi \rangle = \frac{kT}{2} \tag{3.3}$$

which is the important result that for the harmonic oscillator the thermal meen values of potential and kinetic energy are equal.

Accordingly, the internal energy of one mole is

$$u = 3L\, kT = 3RT \tag{3.4}$$

and specific heat is thus

$$c_V = 3R. \tag{3.5}$$

In a certain "normal" regime of temperature, which includes room temperature, this result is experimentally confirmed as a rule, but with some exceptions. It is called the *Dulong-Petit rule*. An outstanding exception is the diamond, which specifically gave reason to the development of the quantum mechanical theory of specific heat of solids. We shall present two steps of this theory, remarkable in history of physics.

***Einstein's* theory.** In a theory developed by *A. Einstein* in 1907, each molecule of the solid is described as individualized and to be bound to a rest position at its lattice site by a spherical symmetric harmonic force. Dynamically one mole of the solid represents $3L$ independent linear oscillators with the same frequency v. In comparison with this it should be stressed that in the picture which has us led to the *Dulong-Petit* rule, the frequencies of the normal vibrations were allowed to be different. To obtain the specific heat of the solid in the model of *Einstein*, we may return to the result eq. (1.14) for the vibrational heat of the one-dimensional oscillator which we have to multiply by the factor 3. Specific heat c_V as function of T can be taken from Figure 40 if we multiply c_{vibr} by 3. This was the first theory which explained why diamond at room temperature has a value of c_V which is considerably smaller than the *Dulong-Petit* value $3R$. The reason is that room temperature is smaller than Θ, which is relatively high for diamond, a material with very high resistance against deformation, and thus with an especially high frequency v. The theory, however, fails in the limit $T \to 0$. Empirically, c_V goes to zero like T^3 and not like $\exp(-\Theta/T)$, as was to be expected by this theory.

***Debye's* theory.** The last mentioned T^3-law was first explained by *P. Debye* in 1911 by describing the solid to be an continuous elastic matter. The normal frequencies of the system of harmonically coupled molecules are replaced with the frequency spectrum of standing waves in the elastic continuous body. We obtain this spectrum by the same method of counting the standing waves in a cube, as done in chapter 4.1.5. Here it is not adequate to refer to the momentum vector $\vec{P}$ as done with eq. (5.17) in chapter 4.1.5, but to introduce the "wave vector" perpendicular to the wave front:

$$\vec{q} = \frac{1}{\lambda}\vec{\alpha}. \tag{3.6}$$

Eq. (5.19) of chapter 4.1.5 then assumes the form

$$\{n_1, n_2, n_3\} = 2l\vec{q} \tag{3.7}$$

with the result that

$$dz = 8V\,d^3q \tag{3.8}$$

is the number of eigenfrequencies of the standing waves associated with the volume element d^3q of $\vec{q}$-space. If γ is the velocity of sound, q is connected with the frequency v by

$$q = \frac{v}{\gamma}. \tag{3.9}$$

In a solid medium, however, there are not only longitudinal waves like in air, but there are moreover transversal waves and we have to distinguish between a longitudinal sound velocity γ_L and a transversal one γ_T. Each wave vector $\vec{q}$ is associated with one longitudinal and two transversal waves. The description of the solid to be a continuous medium is limited to linear extensions of the order of the lattice constant of the crystal. Therefore we restrict the frequencies just up to a finite cut-off value ν_0. By integration over the positive octant of the $\vec{q}$-space we then obtain

$$\int d^3 q \ldots = \frac{4\pi}{8} \int_0^{q_0} dq\, q^2 \ldots = \frac{4\pi}{8} \frac{1}{\gamma^3} \int_0^{\nu_0} d\nu\, \nu^2 \ldots . \tag{3.10}$$

The cut-off is chosen so that the number of linear oscillators is equal to the number $3L$ of degrees of freedom of the L mass points which represent the molecules in our picture:

$$\nu\, 4\pi(\gamma_L^{-3} + 2\gamma_T^{-3})\, \nu_0^3 = 3L. \tag{3.11}$$

The order of the lattice constant a is connected with the mole volume ν by

$$\nu = a^3 L. \tag{3.12}$$

Comparison of a with the corresponding expression in eq. (3.11) shows that an adequately chosen intermediate value λ_0 between the longitudinal and transversal wave length indeed is of the order a, according to

$$\frac{4\pi}{3} \left(\frac{a}{\lambda_0}\right)^3 = 1. \tag{3.13}$$

Each standing wave with frequency ν dynamically represents a linear oscillator and contributes to the internal energy of the whole body with the amount of eq. (1.10)

$$\langle \epsilon \rangle = \left(\frac{1}{e^\alpha - 1} + \frac{1}{2}\right) h\nu \tag{3.14}$$

with

$$\alpha = \frac{h\nu}{kT}, \tag{3.15}$$

(not to be mistaken for $\vec{\alpha}$ in eq. (3.6)). Thus the internal energy of one mole is

$$u = \frac{9L}{\nu_0^3} \int_0^{\nu_0} d\nu\, \nu^2 \left[\left(\exp\frac{h\nu}{kT} - 1\right)^{-1} + \frac{1}{2}\right] h\nu. \tag{3.16}$$

With the introduction of the so-called *Debye temperature* T_D as a characteristic quantity of the material in

$$\alpha_0 = \frac{h\nu_0}{kT} = \frac{T_D}{T} \tag{3.17}$$

and with the function

$$\Psi(\alpha_0) = \alpha_0^{-3} \int_0^{\alpha_0} d\alpha\, \frac{\alpha^3}{e^\alpha - 1} \tag{3.18}$$

we obtain

$$u = 9RT \, \Psi \left(\frac{T_D}{T} \right). \tag{3.19}$$

The *Debye* temperature T_D is obtained by fitting the numerically calculated values

$$c_V = \frac{\partial u}{\partial T} \tag{3.20}$$

to the observed values of the specific heat. In Figure 42 c_V is plotted with a solid line as a function of T. For comparison, the result of the *Einstein* theory with Θ equal to the *Debye* temperature T_D is represented as the broken line. Whereas T_D is quite below room temperature for most substances, corresponding to the *Dulong-Petit* rule, diamond has the exceptional high value of 1860 °K.

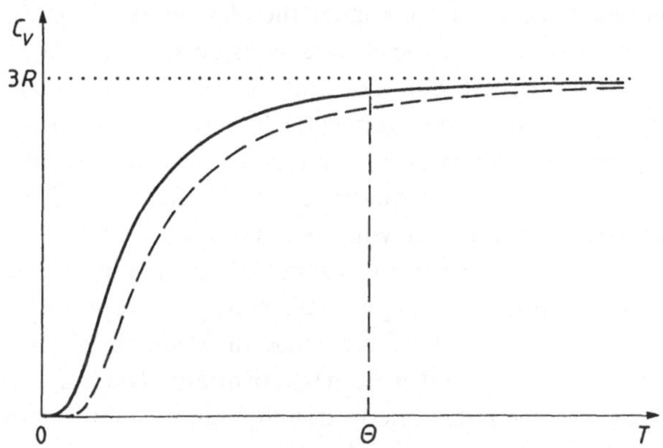

Figure 42 Specific heat c_V of a solid as a function of temperature T corresponding to the *Debye* theory (solid line) and the *Einstein* theory (broken line). For comparison, the *Debye* temperature T_D and the vibration temperature Θ of *Einstein* are assumed to be equal.

We are interested in the behavior of c_V for the limiting process that temperature T goes to zero, because here *Einstein*'s theory was inadequate. In this limit we may write

$$\Psi \left(\frac{T_D}{T} \right) = \left(\frac{T}{T_D} \right)^3 \int_0^\infty d\alpha \, \alpha^3 \sum_{n=1}^\infty e^{-n\alpha} = 6 \sum_{n=1}^\infty n^{-4} = \frac{\pi^4}{15}, \tag{3.21}$$

and obtain

$$c_V = \frac{12}{5} \, \pi^4 R \left(\frac{T}{T_D} \right)^3. \tag{3.22}$$

This yields the observed T^3-law. Moreover, it allows us to find the *Debye* temperature T_D directly. The values for different substances obtained in this way give relatively good results for c_V, with eqs. (3.19, 3.20), with regard to the whole regime of temperature.

4.3   Magnetism

In this subsection we shall discuss thermodynamic properties of magnetic substances on the basis of a microscopic theory. For these systems essential features can be explained by relatively simple microscopic models. Therefore, this is one of the most successful fields of application in statistical thermodynamics.

Roughly we can distinguish three types of magnetism: diamagnetism, paramagnetism and ferromagnetism. The first one is explained by the phenomenon that a magnetic field induces a magnetic moment in each molecule which lasts only so long as the field is present. Para- and ferromagnetism are caused by the orientation of already existing elementary magnetic moments. In the absence of the external magnetic field, they may be without a mean orientation. Then the substance is not magnetized. The elementary magnetic moments are those of molecules atoms or of electrons. In the ordinary ferromagnet they are always electrons. The essential difference between para- and ferromagnets is the following. In the ferromagnet the elementary magnets are correlated with one another. The orientation of each one is essentially influenced by the others. Therefore, a common mean orientation can arise without an external magnetic field and last for an arbitrary long time. The ferromagnet then has a permanent magnetization. In the ideal paramagnet the elementary magnets are not correlated with one another with respect to their orientation. A common mean orientation will last only in the presence of an external field. Otherwise it will soon be destroyed by the thermal motion. This separation into the three types of magnetism is an idealization which, however, is very useful for understanding the origin of the magnetic properties of matter. Diamagnetism has to be studied as a property of the single molecule and is not the subject of a statistical theory, whereas para- and ferromagnetism are. These are essentially determined by the counterplay between the orientation by the magnetic field and the disarray by heat.

We shall describe the magnet to be a system of elementary magnets. By convention, the vector sum of their magnetic moments in a unit volume is defined as the *magnetization* $\vec{M}$. Therefore we shall consider not one mole of the substance but the system of all elementary magnets in a unit volume as our "magnet". The interaction between this system and a homogeneous magnetic field $\vec{B}$ is

$$\mathbf{H}_1 = -\vec{B}\vec{M}. \tag{1}$$

In general there also is an interior energy $\mathbf{H}_0$ of the interaction between the elementary magnets. The total energy of the system then is

$$\mathbf{H} = \mathbf{H}_0 + \mathbf{H}_1. \tag{2}$$

To avoid a confusion between different descriptions, it seems worthwhile to point out that the probability distribution in phase space

$$\rho = \exp\left[\beta(\Phi - \mathbf{H}_0 + \vec{B}\vec{M})\right] \tag{3}$$

can be interpreted in different ways. We can identfy $\langle H_0 \rangle$ with the internal energy of the magnet and interprete ρ as a magnetic field ensemble of type eq. (3.8) of chapter 2.2.3. Φ then is identical with the corresponding *Gibbs* free energy G and $\langle H \rangle$ with enthalpy. Another possibility is that we identify $\langle H \rangle$ with the internal energy, and interprete ρ as a canonical distribution. Then Φ is identical with the *Helmholtz* free energy F. In both cases the corresponding partition function is

$$Z = \text{tr} \exp [\beta (\vec{B} \vec{M} - H_0)] \tag{4}$$

and with it

$$\langle \vec{M} \rangle = \frac{1}{\beta} \frac{1}{Z} \left(\frac{\partial Z}{\partial \vec{B}} \right)_\beta , \tag{5}$$

$$\langle H \rangle = - \frac{1}{Z} \left(\frac{\partial Z}{\partial \beta} \right)_{\vec{B}} . \tag{6}$$

These two equations will be important in the following chapters.

4.3.1 Paramagnetism

The ideal paramagnet contains many elementary permanent magnets in the form of molecules or electrons which do not influence one another. If no external magnetic field is present, the elementary magnets do not prefer any direction in space. In thermal equilibrium, the resulting magnetization will be zero. Any prevalence of a direction will fast be destroyed by the irregular thermal perturbations. The macroscopic magnet thus has no permanent magnetization. As the elementary magnets are not correlated with one another, it suffices to consider only a single one with respect to the probability distribution over the microstates. In a paramagnetic gas the elementary magnet is a molecule. In some paramagnetic substances it is a single electron. In rare earth elements it is an atom. In any case we can put the elementary magnetic moment $\vec{\mu}$ proportional to the elementary angular momentum $\hbar \vec{j}$ with the quantum number j:

$$\vec{\mu} = g \vec{j}. \tag{1.1}$$

We assume that the magnetic field $\vec{B}$ is homogeneous in the extension of the whole magnet The component of the angular momentum in the direction of the magnetic field is $\hbar m$ with the $2j + 1$ possible values

$$m = j, j - 1, ..., -j \tag{1.2}$$

of the so called *magnetic quantum number m*. We write the energy of the elementary magnet in the magnetic field in the form

$$E_m = - \vec{\mu} \vec{B} = - \frac{m}{j} \mu B. \tag{1.3}$$

As the energy levels E_m are not degenerate, the canonical distribution of the magnet is given by the probability

$$P_m = \frac{1}{Z} \exp (- \beta E_m) = \frac{1}{Z} \exp \left(\frac{m}{j} \alpha \right) \tag{1.4}$$

of the orientation with the quantum number m. In this expression the abbreviation

$$\alpha = \frac{\mu B}{kT} \tag{1.5}$$

is used. The partition function is

$$Z = \sum_{m=-j}^{j} \exp\left(-\beta E_m\right) = \sum_{m} \exp\frac{\alpha m}{j}. \tag{1.6}$$

On account of eq. (5) of the introduction to the subsection 4.3 we obtain

$$\frac{1}{j}\langle m\rangle = \frac{1}{Z}\frac{\partial Z}{\partial \alpha}. \tag{1.7}$$

With the further abbreviation

$$a = \exp\frac{\alpha}{j} \tag{1.8}$$

we write

$$Z = a^{-j}\sum_{s=0}^{2j} a^s = \frac{a^{j+1/2} - a^{-j-1/2}}{a^{1/2} - a^{-1/2}} \tag{1.9}$$

$$Z = \frac{\sinh\left[\alpha(2j+1)/2j\right]}{\sinh\left(\alpha/2j\right)}, \tag{1.10}$$

$$\frac{1}{j}\langle m\rangle = \frac{2j+1}{2j}\coth\left(\frac{2j+1}{2j}\alpha\right) - \frac{1}{2}\coth\frac{\alpha}{2j} \equiv L_j(\alpha). \tag{1.11}$$

This function is called the *Brillouin function* $L_j(\alpha)$. We obtain the magnetization which is parallel to $\vec{B}$

$$M = N\mu L_j\left(\frac{\mu B}{kT}\right), \tag{1.12}$$

where μ is the absolute value of the magnetic moment of one elementary magnet and N is the number of these magnets in unit volume. For electrons, and more generally for spin $j = 1/2$, we obtain the so-called *Langevin function*

$$L_{1/2}(\alpha) = \tanh\alpha. \tag{1.13}$$

The limiting process that j goes to infinity leads us to the case of classical physics that the angle between $\vec{\mu}$ and $\vec{B}$ can assume continuous values. It yields the "classical" *Brillouin* function

$$L_\infty(\alpha) = \coth\alpha - \frac{1}{\alpha}. \tag{1.14}$$

These designations are not unique. Sometimes all L_j are called *Langevin* functions. They are plotted in Figure 43 for $j = 1/2$ and for $j = \infty$. The curces for all other values of j lie between these two lines Their steepness decreases with increasing j.

For $\alpha \to \infty$ all L_j go to 1, which means that all elementary magnets assume the direction of the field. This however, is not the usually observed situation. On the contrary, in standard cases the regime of small values α is observed for which L_j is practically

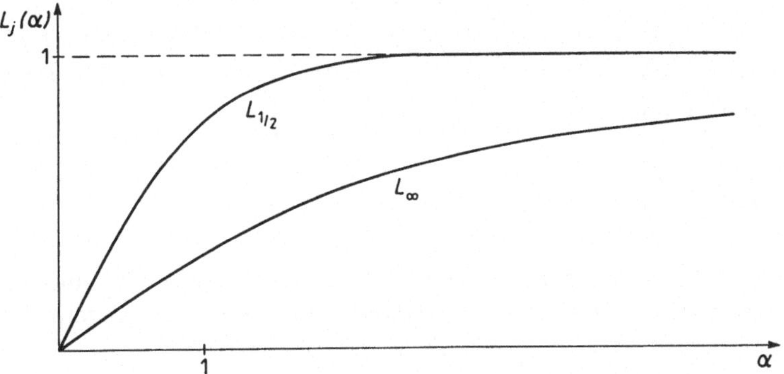

Figure 43 The *Langevin* function $L_{1/2}$ and the classical *Brillouin* function L_∞

linear with respect to α. The connection between magnetization and field then is linear as well

$$M = \chi B, \tag{1.15}$$

where χ is called the *magnetic susceptibility* of the material. The derivative of L_j for vanishing α is

$$\left(\frac{\partial L_j}{\partial \alpha}\right)_{\alpha=0} = \frac{1}{3}\left(1 + \frac{1}{j}\right). \tag{1.16}$$

This yields the so-called *Curie law*

$$\chi = \frac{C}{T}. \tag{1.17}$$

According to experience, this law holds in a very large regime of T and B. Deviations are found only in extremely strong fields and at low temperatures. Due to eq. (1.16) the *Curie constant* is

$$C = \left(1 + \frac{1}{j}\right)\frac{\mu^2}{3k}. \tag{1.18}$$

As Figure 43 shows, the deviations from the *Curie* law are to be expected first with increasing α if $j = 1/2$.

4.3.2 *Weiss* Theory of Ferromagnetism

Ferromagnetism is essentially determined by the interaction and the correlations between the elementary magnets in the material. *P. Weiss* proposed a relatively simple model in which the influence on one elementary magnet of all the others is simulated by an internal magnetic field which itself is proportional to the total magnetization $\vec{M}$ of the material. Thus, if $\vec{B}$ is the external magnetic field, the effective field acting on the single elementary magnet is

$$\vec{B}' = \vec{B} + W\vec{M}. \tag{2.1}$$

The proportionality constant W is called the *Weiss factor*. The internal field $W\vec{M}$ is called the *mean field* of the elementary magnets, or the *molecular field*. Each elementary magnet then can be dealt with separately, like in the paramagnet, if $\vec{B}$ is replaced with $\vec{B}'$. The elementary magnets are electrons which have the spin 1/2. It is assumed that $\vec{M}$ is parallel to $\vec{B}$. The result of eq. (1.12) for $j = 1/2$ with eq. (1.13) has to be changed to

$$M = M_\infty \tanh\left[\frac{\mu}{kT}(B + WM)\right]. \tag{2.2}$$

This equation is the essential starting point of the *Weiss* theory and became very famous because it represented the first microscopic model which was able to describe the phenomenon of a phase transition.

For the discussion of this equation, we rewrite it in a convenient form by introducing the dimensionless quantities

$$\eta = \frac{M}{M_\infty}, \tag{2.3}$$

$$\alpha = \frac{\mu}{kT}(B + WM), \tag{2.4}$$

$$a = \frac{B}{WM_\infty}, \tag{2.5}$$

$$t = \frac{kT}{\mu WM_\infty} = \frac{T}{T_c}. \tag{2.6}$$

Eq. (2.2) then can be replaced by the two equations

$$\eta = \tanh\alpha, \tag{2.7}$$

$$\alpha = \frac{a + \alpha}{t}. \tag{2.8}$$

A graphical method of solving these equations shows the character of the solutions very directly. In Figure 44 the function η of α and the straight line eq. (2.8) are plotted in the (α, η)-plane. The intersection of the two lines yields the solution η to a given value t.

We are particularly interested in the case of the vanishing external field B, i.e. that a is zero. In this case we distinguish two regimes of t. For $t < 1$, that means for T below the *critical temperature* T_c which corresponds to $t = 1$, the quantity η, and thus magnetization M, does not vanish. That is to say, there is a so-called *spontaneous magnetization* also in absence of an external field. It is represented in Figure 45 as a function of temperature. The spatial direction of the spontaneous magnetization is casual. It is determined by the casual majority of elementary magnets when the equilibrium started to develop.

For $T > T_c$ the magnetization is zero. It should be mentioned that a solution $M = 0$ exists also for $T < T_c$. It is, however, unstable, as the thermodynamic stability criteria would show which we discussed in the chapter 2.3.8. Here we only state that this solution for $T < T_c$ does not correspond to a realizable equilibrium.

At the critical temperature T_c a "phase transition of second order" takes place.

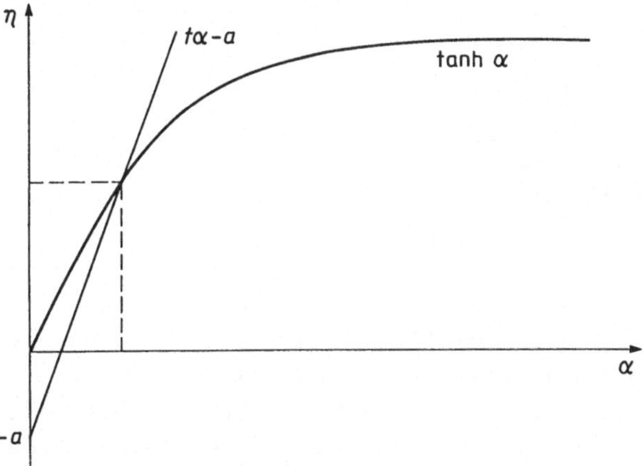

Figure 44 The *Weiss* theory. Graphical solution of the equations for the reduced magnetization η

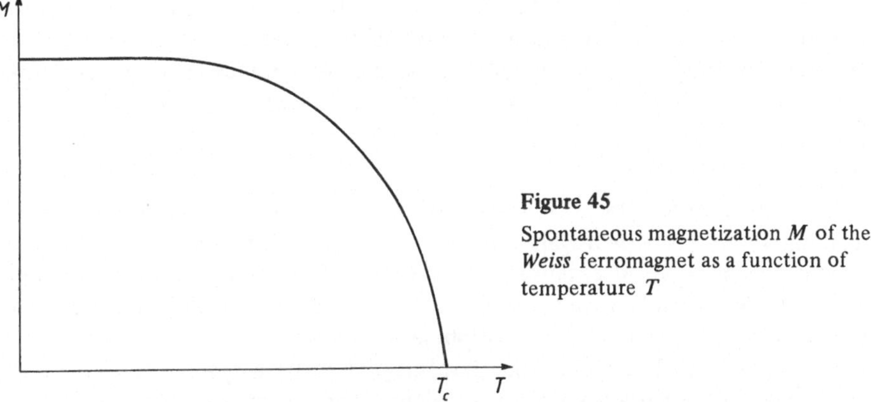

Figure 45

Spontaneous magnetization M of the *Weiss* ferromagnet as a function of temperature T

The spontaneous magnetization, which is finite below this point, vanishes above it. For $T < T_c$ two "phases" are possible. These are the two states of finite magnetization with opposite direction. The change from one to the other state is a "phase transition of first order".

Now we shall consider the behavior of the material in the temperature regime $T > T_c$ and for a nonzero external field B. We shall restrict the consideration to small fields, this means that a is small. As Figure 44 shows, α then is small as well. Thus eq. (2.7) reduces to

$$\eta = \alpha \tag{2.9}$$

yielding

$$\eta = \frac{a}{t-1}, \tag{2.10}$$

or explicity

$$M = \frac{C}{T - T_c} B, \tag{2.11}$$

which by extension of the designation of the law eq. (1.17) is also called a *Curie law*. The corresponding *Curie constant* is

$$C = \frac{\mu}{k} M_\infty. \tag{2.12}$$

We see, that in the regime $T > T_c$ the model behaves like a paramagnet with a susceptibility proportional to $(T - T_c)^{-\gamma}$ with an exponent γ equal to one.

These results indeed correspond to the behavior of ferromagnetic substances in a roughly qualitative way if we do not take account of hysteresis effects and other detailed properties of different ferromagnetic materials.

More detailed models show an exponent γ different from 1 and dependent on the dimension in space. In the study of such models not only the singularity of the susceptibility at the critical temperature can be described by a so-called "critical exponent" which is defined as the exponent of $(T - T_c)$ for the limit of $T \to T_c$. Such critical exponents are defined for different quantities, like for the "order parameter" M, or specific heat. The investigation of universality classes of systems with common critical exponents became a very extended field in the last decades in which particular methods of analysis were developped, the so-called "renormalization group" technics.

*4.3.3   The *Ising* Model

Since about 1925 a model for the ferromagnet is discussed which, notwithstanding that it is a relatively simple and rough simulation, was very successful and became the basis of an extended theoretical analysis which is not yet completed. This so-called *Ising model* proved useful not only with respect to ferromagnetism, but also with respect to the general theory of phase transitions.

The model describes the ferromagnet to be a system of elementary magnets which can only assume two possible orientations in opposite directions in space. These magnets are placed on the sites of a regular crystal lattice.

The quantum mechanical spin in the nonrelativistic description of the electron, the so-called *Pauli spin* can assume only two directions in a given magnetic field as eigenstates of the component in the field axis, parallel or antiparallel to the field. These two eigenstates form a complete orthogonal basis of the state space of the spin of the single electron. This means that the manifold of all pure spin states of the electron is the two-dimensional vector space of all normalized linear combinations of the two basic states. This manifold corresponds to the fact that the field axis can be any in space. The system of such *Pauli* spins on the sites of a crystal lattice is called the *Heisenberg model* of the ferromagnet. We mentioned this, to make clear which approximation is made with the *Ising* model in comparison to the more complete description by the *Pauli* spin. The restricted two states which the spin is only able to assume in the *Ising* model, may be characterized by a spin variable s with only two possible values: with $+1$ associated with the "up" state, and with -1 associated with the "down" state.

Each spin in the lattice may have γ nearest neighbors and we assume that there is interaction only between nearest neighbors. We again consider a lattice of all N spins fulfilling a unit volume, because the vector sum of their magnetic moments is the magnetization of the material. A microstate of this system is described by an ordered set of N values of the spin variables

$$\{s_1, s_2, \ldots, s_N\} \equiv \{s_i\} \tag{3.1}$$

each s_i being either $+1$ or -1.

The interaction energy between a pair of neighboring spins (i, j) is assumed to be proportional to the product $s_i s_j$. With the magnetic moment μ of the single electron, the total energy in the microstate of the whole system of the N spins is

$$H\{s_i\} = -\mu B \sum_{i=1}^{N} s_i - \epsilon \sum_{[ij]} s_i s_j, \tag{3.2}$$

where the second summation runs over pairs of nearest neighbors only. This is designated by the bracket $[\ldots]$. The multiplier ϵ is positive in a ferromagnet. There are, however, also so-called *antiferromagnets* with negative ϵ. In the following we consider ferromagnets only.

The partition function of the canonical distribution is

$$Z(T, B) = \sum_{\{s_i\}} \exp(-\beta H\{s_i\}). \tag{3.3}$$

It gives *Helmholtz* free energy F, internal energy U, and magnetization M in the relations

$$F = -\beta \ln Z, \tag{3.4}$$

$$U = -\frac{\partial}{\partial \beta} \ln Z, \tag{3.5}$$

$$M = \frac{1}{\beta} \frac{\partial}{\partial B} \ln Z. \tag{3.6}$$

The number N is assumed to be so large that we may treat the spins situated at the surface of the lattice like those in the bulk. In a given microstate $\{s_i\}$ we have N_+ up-spins, N_- down-spins, N_{++} neighboring up-pairs, N_{--} neighboring down-pairs, and N_{+-} neighboring up-down-pairs. The number N_{+-} of up-down and N_{-+} of down-up-pairs is not discriminated and thus equal. All these numbers are not independent of one another. To find relations between them, we refer to Figure 46 and make the following construction. Let lines be drawn outgoing from each up-spin to all nearest neighbors. Then lines between up-pairs become twofold, those between up-down-pairs simple. The total number of lines in the lattice is

$$\gamma N_+ = 2N_{++} + N_{+-}. \tag{3.7}$$

This relation holds also if all spins are converted into the opposite direction:

$$\gamma N_- = 2N_{--} + N_{+-}. \tag{3.8}$$

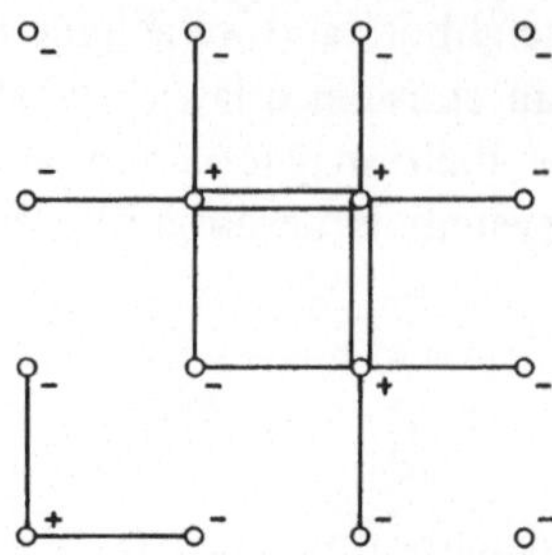

Figure 46

The *Ising* model. Method of counting the different pairs in a lattice. Lines are drawn outgoing from each up-spin (+) to all next neighbours. Lines between up-up-pairs become twofold, between up-down pairs remain simple. Between down-down pairs there are no lines.

Moreover, the relation

$$N = N_+ + N_-$$ (3.9)

holds. With this we are able to eliminate all the numbers with down-spins and we obtain

$$\sum_i s_i = N_+ - N_- = 2N_+ - N,$$ (3.10)

$$\sum_{[ij]} s_i s_j = N_{++} + N_{--} - N_{+-} = 4N_{++} - 2\gamma N_+ + \frac{\gamma}{2} N.$$ (3.11)

It is convenient to introduce two independent parameters with

$$N_+ = \frac{1+\eta}{2} N,$$ (3.12)

$$N_{++} = \frac{1+\sigma}{2} \frac{\gamma}{2} N$$ (3.13)

varying in the interval between -1 and $+1$. The parameter η characterizes the total sum of all spins

$$\sum_i s_i = N\eta,$$ (3.14)

and thus the magnetization if here and in the following we put μ equal to one, with adequate choice of units. The parameter σ characterizes the surroundings formed by the nearest neighbors of the single elementary magnet. η is called the *long range order parameter*, τ the *short range order parameter*. We may write

$$\sum_{[i,j]} s_i s_j = \gamma N \left(\sigma - \eta + \frac{1}{2} \right),$$ (3.15)

and we obtain for the total energy of the microstate

$$H\{s_i\} = -BN\eta - \epsilon\gamma N \left(\sigma - \eta + \frac{1}{2} \right).$$ (3.16)

One pair (η, σ) is associated with many microstates. These are all microstates with the same numbers N_+, N_{++}. The number of these microstates shall be designated by $g(\eta, \sigma)$. Then we may write

$$Z = \sum_{\sigma\eta} g(\eta, \sigma) \exp\left\{\beta N\left[\mu B\eta + \epsilon\gamma\left(\sigma - \eta + \frac{1}{2}\right)\right]\right\}.$$

(3.17)

In principle this expression finally determines

$$M = N\mu\langle\eta\rangle = M_\infty\langle\eta\rangle,$$

(3.18)

$$U = MB + N\epsilon\gamma\left\langle\sigma - \eta + \frac{1}{2}\right\rangle$$

(3.19)

The evaluation of these relations and generally the analysis of the *Ising* model has been the subject of extended research which is not yet finished. Rigorous solutions of the quoted relations were found only for the dimension one and two. The case of dimension one doesn't show a phase transition. The case of dimension two already is very complicated. It was solved by *L. Onsager* in 1944. Many work, also for higher dimensions than three, is done by computer calculation. For dimension three a power expansion of the partition function with respect to β was very useful. It is called a "high-temperature expansion" because it converges the faster, the higher the temperature. Computer calculations were possible also of expansion coefficients of higher order and gave good results over a large temperature regime.

In the following two chapters we shall present two different approximations which were carried out alread relatively early and which became very renowned because they separated the different influences of the so-called short range order from the long range order.

*4.3.4 The Long Range Order

In 1934 for the *Ising* model an approximation, in which the correlations between the nearest neighbors were suppressed, was developed by *W. L. Bragg* and *E. J. Williams*. This was done by putting the probability that neighboring spins are in the same state equal to the product of the single probabilities of being in this state. In particular:

$$P_{++} = P_+ P_+.$$

(4.1)

Corresponding to eqs. (3.12, 3.13) we again write

$$P_+ = N_+/N = \frac{1 + \eta}{2},$$

(4.2)

$$P_{++} = 2N_{++}/\gamma N = \frac{1 + \sigma}{2}.$$

(4.3)

The approximation of eq. (4.1) then reads

$$\frac{1 + \sigma}{2} = \left(\frac{1 + \eta}{2}\right)^2.$$

(4.4)

Thus in the partition function Z of eq. (3.17)

$$\sigma - \eta + \frac{1}{2} = \frac{\eta^2}{2} . \tag{4.5}$$

The number $g(\eta, \sigma)$ of microstates to given N_+, N_{++} now depends on N_+ alone:

$$g = \binom{N}{N_+} = \frac{N!}{N_+! \, N_-!} . \tag{4.6}$$

We assume that all numbers occurring in this expression are large enough to apply the *Stirling* formula

$$\ln N! \approx N \ln N - N \tag{4.7}$$

if they really contribute noticeable to Z, and we obtain

$$\ln g = -\frac{N}{2}(1+\eta)\ln\frac{1}{2}(1+\eta) - \frac{N}{2}(1-\eta)\ln\frac{1}{2}(1-\eta), \tag{4.8}$$

$$Z = \sum_\eta \exp\left\{\frac{N}{2}\left[-(1+\eta)\ln\frac{1}{2}(1+\eta)-(1-\eta)\ln\frac{1}{2}(1-\eta)+(2\mu B+\epsilon\gamma\eta^2)\right]\right\}. \tag{4.9}$$

As the total number N of all spins is a very large number, say of the order of 10^{20}, the terms in the sum in eq. (4.9) have an extremely sharp maximum. Its value so overwhelms the other terms that it is allowed to replace the whole sum with its largest term. This is the same approximation which was already introduced as the *saddle point* method in chapter 2.2.4. We have to find the largest term in the sum. For this the derivative of the exponent in eq. (4.9) with respect to η vanishes. This yields:

$$\ln\frac{1+\eta}{1-\eta} = 2\beta(\mu B + \epsilon\gamma\eta). \tag{4.10}$$

We can write this equation by introducing

$$\alpha = \beta(\mu B + \epsilon\gamma\eta) \tag{4.11}$$

in the form of the two equations

$$\eta = \tanh\alpha, \tag{4.12}$$

$$\alpha = \frac{1}{t}(a + \eta), \tag{4.13}$$

which already occurred in the *Weiss* theory eqs. (2.7, 2.8). Therefore we can return to this theory with the identification

$$\frac{\epsilon\gamma}{M_\infty} = \frac{\epsilon\gamma}{\mu N} = W. \tag{4.14}$$

With respect to the thermal equation of state, the result is identical to that of the *Weiss* theory. The more detailed model we are discussing now, however, also gives information about the caloric equation of state. That is to say, it allows us to calculate the specific heat. For this purpose we consider the case that no external field B is present.

The internal energy is

$$U = -N \frac{\gamma}{2} \epsilon \eta^2,\tag{4.15}$$

and the specific heat related to unit volume is

$$c = \frac{\partial U}{\partial T} = -N\gamma\epsilon\eta \frac{\partial \eta}{\partial T}.\tag{4.16}$$

η can be obtained by solving eqs. (4.12, 4.13) as for the *Weiss* model. With this result then eq. (4.16) yields c as a function of temperature T. This function is plotted in Figure 47 and can easily be understood by the slope of $\eta(T)$ in Figure 45. The specific heat makes a finite jump at the critical temperature T_c and vanishes for higher temperatures. Real ferromagnets also show a singularity at the critical temperature which in broad outline is not totally different. In details, however, there are differences which will be discussed in the next chapter.

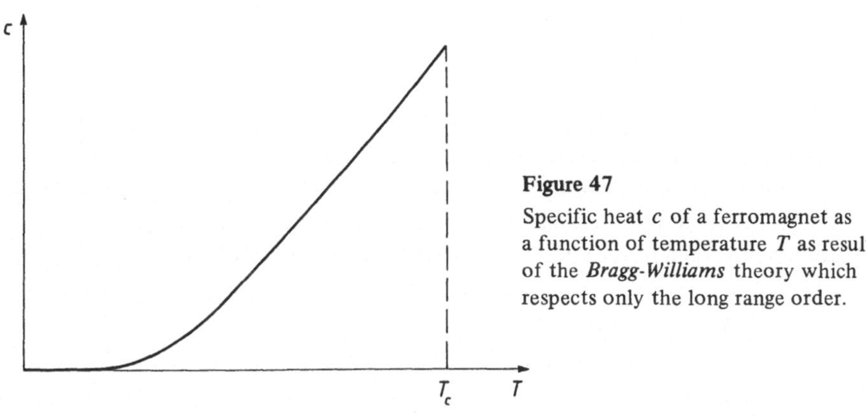

Figure 47

Specific heat c of a ferromagnet as a function of temperature T as result of the *Bragg-Williams* theory which respects only the long range order.

It seems worthwhile to point out a particular property of this system which is characteristic for systems with a phase transition. This is that the intensities (here only T) do not determine the thermal state uniquely. In our case, for zero magnetic field and $T < T_c$ two different states with opposite values of the magnetization M are possible at the same temperature, the two *phases*. We obtained these two values as the M-values of two maxima of the probability distribution. The mean value of M in this distribution, however, is zero, yet not in each one of the two phases. In such a case, the mean value of the extensity M in this distribution is not identified with the macroscopic variable. This distribution would correspond to the situation that the sign of M and thus the "phase", in which the substance is, is unknown. If, however, the sign is known, the probability distribution will be replaced by the one-peak distribution with only one maximum at M of this sign. The other peak will be omitted. The sign of M and the one-peak distribution is determined if a nonzero field B is given, however small it may be. We should obtain the one-peak distribution automatically in an analytic calculation by the adequate sequence of limiting processes; first performing the limit for N going to infinity, and afterwards the limit for B going to zero. The phenomenon that for vanishing B unlike to the unmagnetized state for $T < T_c$, the ferromagnetic state is not symmetric in ordinary space is typical for phase transitions and is called the *spontaneous symmetry breaking*.

*4.3.5 The *Bethe-Peierls* Approximation

We shall now consider an approximation which, unlike to the *Bragg-Williams* approximation, does not disregard the short range correlations. The method was first developed in 1935 by *H. Bethe* for order-disorder phenemona in alloys[16]. *R. Peierls* applied it to the *Ising* model in 1936[17]. In this approximation correlations between two spins are respected, but not correlations between more than two spins. We shall be concerned here only with the case that no external field is present.

Correlations between neighboring pairs are introduced into the description with the probability $P(s, n)$ that a given spin has the value s and that, simultaneously, exactly n of its γ nearest neighbors have spin $+1$. This probability is assumed to be

$$P(s, n) = \frac{1}{q} \binom{\gamma}{n} \zeta^n \exp\left[-\beta \omega(s, n)\right], \tag{5.1}$$

where

$$\omega(s, n) = -\epsilon s\left[n - (\gamma - n)\right] = -\epsilon s(2n - \gamma) \tag{5.2}$$

is the interaction energy between the "central" spin s and its γ nearest neighbors. In this distribution, n takes the place of the "particle number" of a grand canonical ensemble. We can interprete this assumption as the result of the unbiased guess if the mean value of the energy ω and the mean value of n are known. Correspondingly, ζ is the fugacity of the nearest neighbors. The "grand partition function" of this particular distribution is

$$q = \sum_s \exp\left(-\beta s \gamma \epsilon\right) \sum_{n=0}^{\gamma} \binom{\gamma}{n} \zeta^n \exp\left(2\beta \epsilon s n\right). \tag{5.3}$$

In the expression

$$q = \sum_s \left[\exp\left(-\beta s \epsilon\right) + \zeta \exp\left(\beta s \epsilon\right)\right]^\gamma, \tag{5.4}$$

the summation runs but over the two values $+1$ and -1 of s. Now we can obtain

$$\langle n \rangle = \sum_{sn} n P(s, n) = \frac{\zeta}{q} \frac{\partial q}{\partial \zeta}, \tag{5.5}$$

$$\frac{\langle N_+ \rangle}{N} = \frac{1 + \eta}{2} = P_+ = \sum_n P(+1, n), \tag{5.6}$$

$$\frac{2 \langle N_{++} \rangle}{\gamma N} = \frac{1 + \sigma}{2} = \frac{1}{\gamma} \sum_n n P(+1, n). \tag{5.7}$$

---

[16] *H. A. Bethe*, Proc. R. Soc. (A) **150**, 552 (1935).

[17] *P. Peierls*, Proc. R. Soc. (A) **154**, 207 (1936).
P. R. Weiss, Phys. Rev. **74**, 1493 (1948).

n always runs from 0 to γ. Up to now the quantities in this relations are still indetermined because we do not know the mean value of n, or the fugacity, respectively. We have, however, to respect the self-consistency condition

$$\frac{\langle N_+\rangle}{N} = \frac{\langle n\rangle}{\gamma}. \tag{5.8}$$

It represents a connection between long and short range order and establishes the sought determination.

For the evaluation of these relations we use the abbreviation

$$\exp(\beta\epsilon) = a \tag{5.9}$$

and obtain

$$q = (a^{-1} + \zeta a)^\gamma + (a + \zeta a^{-1})^\gamma, \tag{5.10}$$

$$\frac{\langle n\rangle}{\gamma} = \frac{\zeta}{q}\left[a(a^{-1} + \zeta a)^{\gamma-1} + a^{-1}(a + \zeta a^{-1})^{\gamma-1}\right], \tag{5.11}$$

$$\frac{\langle N_+\rangle}{N} = \frac{1}{q}(a^{-1} + \zeta a)^{\gamma-1}. \tag{5.12}$$

The self consistency relation eq. (5.8) takes the form

$$(1 + \zeta a^2)^\gamma = \zeta\left[a^2(1 + \zeta a^2)^{\gamma-1} + (a^2 + \zeta)^{\gamma-1}\right] \tag{5.13}$$

and yields

$$\zeta = \left(\frac{1 + \zeta a^2}{a^2 + \zeta}\right)^{\gamma-1}. \tag{5.14}$$

a is dependent on temperature. Therefore the last equation determines the function $\zeta(T)$. With this the long and short range order parameters η, σ can be obtained. We use the abbreviation

$$x = \frac{\gamma}{\gamma-1}. \tag{5.15}$$

The results which are obtained after some transformations are

$$\frac{1}{2}(1 + \eta) = \frac{1}{q}a^{-\gamma}(1 + \zeta a^2)^\gamma = (1 + \zeta^{-x})^{-1}, \tag{5.16}$$

$$\frac{1}{2}(1 + \sigma) = \frac{\zeta}{q}a^{2-\gamma}(1 + \zeta a^2)^{\gamma-1} = \zeta a^2\left[(1 + \zeta a^2)(1 + \zeta^{-x})\right]^{-1}, \tag{5.17}$$

and finally

$$\eta = \frac{\zeta^x - 1}{\zeta^x + 1}, \tag{5.18}$$

$$\sigma = \frac{2\zeta a^2}{(1 + \zeta a^2)(1 + \zeta^{-x})} - 1. \tag{5.19}$$

Let us now return to eq. (5.14). We immediately see the following: the solution to zero value of η is $\zeta = 1$. Moreover with ζ also $1/\zeta$ is a solution. If ζ is the solution to η, then $1/\zeta$ is the solution to $-\eta$. Eq. (5.14) can be solved graphically, as demonstrated in Figure 48. In this figure the function

$$f(\zeta) = \left(\frac{1 + \zeta a^2}{a^2 + \zeta} \right)^{\gamma - 1} \tag{5.20}$$

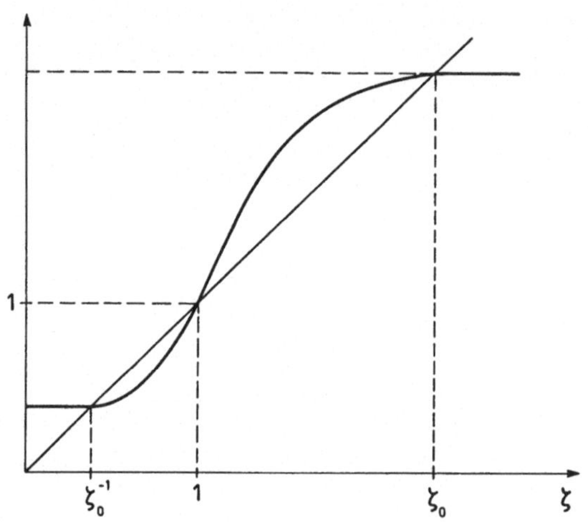

Figure 48

The *Bethe-Peierls* approximation. Graphical solution of the equations for the reduced magnetizaton η.

is represented as a curve. The straight line represents ζ. The intersections of both lines as a rough simulation give for the subcritical temperature regime $T < T_c$ the solutions ζ_0, $1/\zeta_0$ which are stable, as a more detailed analysis would show (the rough simulation is more suited to make the essential points visible than a drawing in correct proportions would do). The third solution $\zeta = 1$ is unstable in this regime. In the supercritical regime $T > T_c$ only the solution $\zeta = 1$ exists and is stable there. The critical temperature T_c corresponds to the special value of a which satisfies

$$f'(1) \equiv (\gamma - 1) \frac{a^2 - 1}{a^2 + 1} = 1, \tag{5.21}$$

yielding

$$kT_c = \frac{2\epsilon}{\ln\left[\gamma/(\gamma - 1)\right]}. \tag{5.22}$$

$\eta(T)$ and with it $M(T)$ is qualitatively of similar shape as that in Figure 45 for the *Weiss* theory and the *Bragg-Williams* approximation.

It is remarkable that the specific heat as a function of temperature behaves differently from the *Bragg-Williams* theory also in its qualitative character. Internal energy is due to eq. (3.15)

$$U = -\epsilon\gamma N\left(\sigma - \eta + \frac{1}{2}\right),\tag{5.23}$$

which yields specific heat related to unit volume by the use of

$$c = \frac{\partial U}{\partial T}.\tag{5.24}$$

The result is plotted qualitatively in Figure 49 as a solid line. The broken line represents the *Bragg-Williams* result. Above the critical temperature, the long range order parameter η vanishes and

$$U = -\epsilon\gamma N\left(\sigma + \frac{1}{2}\right).\tag{5.25}$$

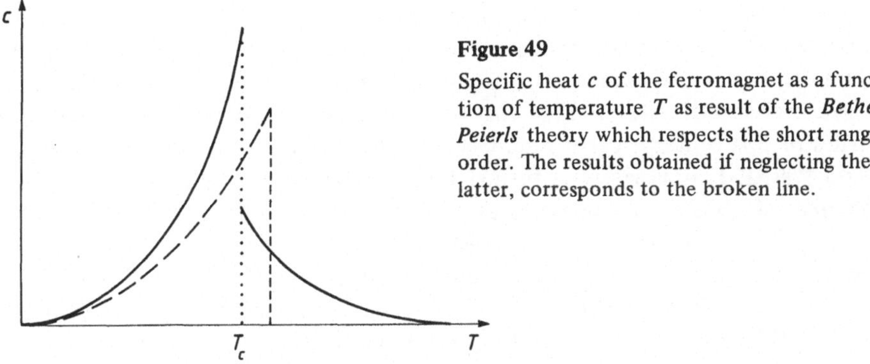

Figure 49

Specific heat c of the ferromagnet as a function of temperature T as result of the *Bethe-Peierls* theory which respects the short range order. The results obtained if neglecting the latter, corresponds to the broken line.

But due to the short range order, which still exists, specific heat is nonzero in agreement with the observation of real ferromagnets.

Subsequent theories on the critical behavior of phase transitions yield divergence of the specific heat of the three-dimensional *Ising* model at the critical point. The high-temperature expansion endorses this.

5   Nonequilibria

Whereas thermodynamics of equilibrium states is an elaborate and closed branche of science, the same cannot be said about thermodynamics of nonequilibria. Of course, thermodynamics in general is restricted to certain macroscopic systems and to certain of their states. Its applicability, however, is not restricted to thermal equilibria. In this section of the book, we shall discuss certain limited extensions of thermodynamics to nonequilibria.

First, it is obvious that the statistical theory of thermal equilibrium includes fluctations deviating from the mean values of observable quantities, the *thermal fluctuations*. The theory of such fluctuations is already a starting point of an extension of thermodynamics to any sufficiently small deviation from the equilibrium. Indeed, this was the way *L. Onsager* developed the nonequilibrium theory of processes which occur sufficiently near an equilibrium to allow the linearization of the macroscopic dynamical equations with respect to the quantities which describe the deviation from the equilibrium. This is the so-called *linear thermodynamics* often simply called *thermodynamics of irreversible processes*, or *TIP*. This theory started in a general systematic form in the thirties of this century, yet it didn't become well known and more extended until after World War II. It is based on a further thermodynamic principle, the *Onsager symmetry principle*, which is of the same rank as the First and Second Law of thermodynamics.

The description of phenomena with thermodynamic means is also possible, however, for states which are farther away from any equilibrium. In this regime we speak of *nonlinear thermodynamics* because there the mentioned linearization would not hold. In this field new and often surprising phenomena occur, like the formation of patterns and the selforganization of structures. It is today the object of very active research, but it has remained the study of many individual properties of different systems and has not yet become a closed theory comparable with that of equilibrium and linear thermodynamics. It is an extended field in its own rights and is not in the scope of this book.

In the first subsection, we shall discuss the fluctuation theory on the basis of the statistical theory of equilibrium. In the second one we shall be concerned with the question how to describe nonequilibrium states by probability distributions in phase space. This will us give the basis for the application of thermal equilibrium quantities, and certain relations between them, to nonequilibria. In the third subsection we shall represent the outlines of linear thermodynamics. The fifth subsection is devoted to the problem of the time scale separation which is a fundamental and necessary condition for any thermodynamic description of nonequilibria at all. It is usually supposed as a property of the "normal" macroscopic systems, yet it has not been deduced in general as a consequence of microscopic dynamics. As this property is often discussed as a fundamental basis, we shall analyse a relatively simple model for which a rigorous analytical treatment is possible. This model will show which steps in the analytical description lead from reversibility to irreversibility and which lead to the time scale separation.

5.1   Thermal Fluctuations

The statistical basis of macroscopic thermodynamics becomes very directly apparent in the phenomenon of thermal fluctuations. These are the deviations of observed values from the thermal mean values. The deviations we speak of are observable by macroscopic means. The probability distributions in phase space allow such deviations from the mean values of those quantities which are used as thermal variables. They, moreover, in principle determine the probability of any fluctuation. This will be the basis of the considerations of this subsection.

The theory of fluctuations represents an important link between different sectors of thermodynamics. This is true in more than one respect. Fluctuations which are observable and describable by macroscopic means can be interpreted in the framework of macroscopic nonequilibrium thermodynamics. On the other hand, in the framework of the statistical theory, fluctuations are deviations from mean values in an equilibrium probability distribution and thus are phenomena of the thermal equilibrium about which only probability statements are possible. So we can say that the fluctuation theory represents a bridge between the macroscopic "deterministic" theory, in which no probabilities occur, and a statistical theory. The phenomenon of fluctuations shows that finally a consistent thermodynamic theory leads us to statistics.

Whereas the thermal equilibrium is not dependent on time, fluctuations are. As they can be understood as equilibrium phenomena, fluctuation theory also represents a bridge between equilibrium and nonequilibrium thermodynamics.

In the first chapter we shall discuss the connection between fluctuations in a generalized canonical distribution and the generalized "susceptibilities". These connections were derived already in chapter 1.3.2 in the framework of general statistics. Now they shall be applied to concrete physical systems.

In chapter 5.1.2 we shall discuss the probability of fluctuations in a given probability distribution in Γ-space with respect to physical applications. This will lead us to the *Einstein fluctuation formula* which attaches probabilities to macroscopically observable events. It was historically the first relation of such character in physics.

The third chapter is devoted to an old problem in thermodynamics. It is the question of whether the existence of a so-called *Maxwellian demon* would be in contradiction to the Second Law or not. As shall be explained, this problem was the reason for *L. Szilard* to detect a connection between macroscopic thermodynamic entropy and the length of a communication, already in 1929, long before the concept of an information measure was developed.

5.1.1   Fluctuations and Susceptibilities

In chapter 1.3.2 it was shown that for a generalized canonical distribution the elements of the correlation matrix

$$Q^{\mu\nu} = \langle \Delta M^\mu \, \Delta M^\nu \rangle \tag{1.1}$$

are equal to the susceptibilities

$$Q^{\mu\nu} = -\frac{\partial M^\mu}{\partial \lambda_\nu}. \tag{1.2}$$

In particular for the diagonal elements with definite ν the equation

$$\langle (\Delta M^\nu)^2 \rangle = -\frac{\partial M^\nu}{\partial \lambda_\nu} \tag{1.3}$$

holds. On the left hand side of this equation, the variance of the fluctuation of M^ν is a measure of how large the fluctuations are on the average. The right hand side is a susceptibility and shows how sensitive the quantity M^ν is with respect to changes of the thermal conjugate. This equation is sometimes called the *static fluctuation-dissipation theorem* with reference to a relation in nonequilibrium thermodynamics which connects fluctuations in a stable stationary state with the response to an external perturbation and is called "fluctuation-dissipation theorem". The name "static" indicates that it is a relation between quantities determined by the equilibrium probability distribution. We already stressed, however, that fluctuations change with time and therefore can be considered as nonequilibrium phenomena as well. We shall now apply eq. (1.3) to special cases.

Fluctuations of the volume. V shall be the volume of a system of particles the number N of which is hold fixed. Then

$$M^\nu = V, \qquad \lambda_\nu = \frac{p}{kT} \tag{1.4}$$

where p is pressure. With the compressibility

$$\kappa = -\frac{1}{V}\left(\frac{\partial V}{\partial p}\right)_T, \tag{1.5}$$

we obtain as a special result of eq. (1.3)

$$\langle (\Delta V)^2 \rangle = kT \kappa V. \tag{1.6}$$

For an ideal gas of N molecules, this yields

$$\left[\frac{\langle (\Delta V)^2 \rangle}{V^2}\right]^{1/2} = N^{-1/2}. \tag{1.7}$$

This means that the relative fluctuations are of the order $N^{-1/2}$. For a macroscopic amount this is extremely small; for one mole around 10^{-11}. The ratio

$$\frac{\langle (\Delta V)^2 \rangle}{V} = kT\kappa = \frac{kT}{p} = \frac{v}{L} \tag{1.8}$$

(with the *Loschmidt* number L) is independent of the amount of matter. It is the gas volume per molecule.

For a real gas the situation becomes different with the approach to the critical point of the gas-liquid phase transition. There the compressibility goes to infinity. The so-called *critical fluctuations* become extremely large in the neighborhood of the critical point. This is responsible for the optical phenomenon of the "critical opalescence" of this system because the local density fluctuations involve fluctuations of the refraction index.

Fluctuations of the particle number. Now we consider the variable number N of particles in a fixed volume V. This is an alternative description of a fluctuating particle density

$$\rho = \frac{N}{V}.$$ (1.9)

By restricting ourselves to small fluctuations we can write

$$\frac{\Delta N}{N} = \frac{\Delta \rho}{\rho}$$ (1.10)

and obtain an equivalent to eq. (1.6) in the form

$$\frac{\langle (\Delta N)^2 \rangle}{N^2} = \frac{kT}{V}\,\kappa.$$ (1.11)

The ratio

$$\frac{\langle (\Delta N)^2 \rangle}{N} = kT\kappa\rho$$ (1.12)

is independent of the amount of matter.

Bose and *Fermi* **gas.** We return to the description of the *Bose* and *Fermi* gas in chapter 4.1.5. We again call n_i the number of particles which are in the pure one-particle state $|i\rangle$ with particle energy ϵ_i. The mean value of this number is in the interaction-free gas

$$\langle n_i \rangle = \{\exp[\beta(\epsilon_i - \mu)] \mp 1\}^{-1}.$$ (1.13)

The upper sign corresponds here and in the following to the *Bose*, the lower sign to the *Fermi* gas. μ is the chemical potential. With

$$\lambda_\nu = -\frac{\mu}{kT},$$ (1.14)

eq. (1.3) now is

$$\langle (\Delta n_i)^2 \rangle = kT \left(\frac{\partial}{\partial \mu} \langle n_i \rangle \right)_T$$ (1.15)

and yields

$$\langle (\Delta n_i)^2 \rangle = \langle n_i \rangle (1 \pm \langle n_i \rangle).$$ (1.16)

In a macroscopic volume, the levels ϵ_i are so dense that only whole bands of many levels can be distinguished from one another. Therefore we ask for the particle number N in an energy band at level ϵ_i with g pure one-particle states. With summation over the levels in the band this is

$$N = \sum^{g} n_i = g n_i.$$ (1.17)

As the particle numbers in different levels are not correlated with one another, the relation holds that

$$\langle (\Delta N)^2 \rangle = \left\langle \sum^{g} \Delta n_i \Delta n_j \right\rangle = g \langle (\Delta n_i)^2 \rangle, \tag{1.18}$$

$$\langle (\Delta N)^2 \rangle = \langle N \rangle (1 \pm g^{-1} \langle N \rangle). \tag{1.19}$$

For small densities this yields the equation

$$\langle (\Delta N)^2 \rangle = \langle N \rangle \tag{1.20}$$

like for the classical ideal gas. The deviation from this equation represents an increase of the fluctuations in the *Bose* gas and a decrease in the *Fermi* gas.

The radiation cavity. We describe the radiation to be a gas of photons. It is a *Bose* gas with vanishing chemical potential μ as was explained in chapter 4.1.6. Thus eq. (1.3) assumes the form

$$\langle n_i \rangle = [\exp(\beta h \nu) - 1]^{-1}, \tag{1.21}$$

where $h \nu$ is the energy ϵ_i of one photon with the frequency ν, corresponding to the one-photon state $|i\rangle$. In the small frequency band between ν and $\nu + \delta \nu$, the number of states $|i\rangle$ is

$$g = V \frac{8\pi}{c^3} \nu^2 \delta \nu, \tag{1.22}$$

as explained in the mentioned chapter (see there eq. (6.5)). The whole energy in this band is

$$E_{\delta \nu} = g n_i h \nu = N h \nu. \tag{1.23}$$

The variance of this energy is due to eq. (1.19):

$$\langle (\Delta E_{\delta \nu})^2 \rangle = \langle (\Delta N)^2 \rangle (h \nu)^2 = (h \nu)^2 N (1 + g^{-1} \langle N \rangle), \tag{1.24}$$

$$\langle (\Delta E_{\delta \nu})^2 \rangle = h \nu \langle E_{\delta \nu} \rangle + g^{-1} \langle E_{\delta \nu} \rangle^2. \tag{1.25}$$

With introduction of the spectral function $f(\nu)$ in

$$\frac{1}{V} \langle E_{\delta \nu} \rangle = f(\nu) \delta \nu \tag{1.26}$$

we obtain

$$\frac{1}{V} \langle (\Delta E_{\delta \nu})^2 \rangle = h \nu \left[1 + \frac{c^3}{8 \pi h \nu^3} f(\nu) \right] f(\nu) \delta \nu, \tag{1.27}$$

where $f(\nu)$ is the expression of the *Planck* radiation formula eq. (6.10) of chapter 4.1.6:

$$f(\nu) = \frac{8\pi h}{c^3} \frac{\nu^3}{\exp(\beta h \nu) - 1}. \tag{1.28}$$

Relation eq. (1.27) had already been given by *A. Einstein* in the year 1909 [18]). *Einstein* stressed the occurrence of the first term 1 in the bracket of the right hand side of eq. (1.27). This term would correspond to photons as independent particles if it alone were present. Classical (non-quantum) theory would only give the second term of the bracket, the "wave term" in which $f(\nu)$ has to be replaced with the classical expression, the limit of vanishing h. This interpretation already uses an interesting discrimination of wave and particle features of light.

5.1.2   Probability of Fluctuations

We shall be concerned with the question of the probability of a definite fluctuation of macroscopic variables in a given equilibrium state ρ^0. The fluctuation may be described by the set V of the deviations V^σ of these variables from their equilibrium values. If we interprete these variables as mean values in the Γ-space, corresponding to the unbiased guess method, they define a distribution ρ over the microstates in the Γ-space, different from ρ^0.

Fluctuations of relative frequencies. We can formulate the problem of finding the probability of this fluctuation as a problem which can be discussed already in the framework of general statistics in the following way. Given a distribution ρ^0 of probabilities P_i^0 over a sample set U_i. What is the probability that in a long series of N independent measurements each result U_i occurs precisely N_i times, that is to say with the relative frequency

$$P_i = \frac{N_i}{N} \tag{2.1}$$

different from P_i^0? We call the occurrence of these relative frequencies the *event R*. It is the sum of many *elementary events X*. Each of these events X is a series of N results with the frequencies N_i but in a special ordered sequence, whereas in R the sequence is arbitrary. The probability of an event X is

$$P(X) = \prod_i (P_i^0)^{N_i}. \tag{2.2}$$

The corresponding bit-number necessary to fix an event X is

$$b(X) = -\ln P(X) = -\sum_i N_i \ln P_i^0. \tag{2.3}$$

This bit-number has to be equal to the sum of the bit-number $b(R)$ needed to fix R plus the bit-number to fix X if R is given:

$$b(X) = b(R) + b(X|R). \tag{2.4}$$

We seek $b(R)$ and obtain it by first determining $b(X|R)$. This can be done in the following way. The bit-number needed to fix one place in a series of N measurements is $\ln N$.

---

[18]) *A. Einstein,* Phys. Zeitschrift **10**, 185 (1909).

The bit-number needed to fix the whole series therefore is $N \ln N$. A permutation of the N_i-times occurring equal results U_i does not change the event X. Therefore the fixing of the sequence of these equal results is redundant for fixing the event X and we have to subtract the corresponding bit-number, obtaining

$$b(X|R) = N \ln N - \sum_i N_i \ln N_i. \tag{2.5}$$

With it we receive

$$b(R) = NK(\boldsymbol{\rho}, \boldsymbol{\rho}^0), \tag{2.6}$$

where

$$K(\boldsymbol{\rho}, \boldsymbol{\rho}^0) = \sum_i P_i \ln \frac{P_i}{P_i^0} \tag{2.7}$$

is the *Kullback* information. Correspondingly the probability of R is

$$P(R) = \exp\left[-NK(\boldsymbol{\rho}, \boldsymbol{\rho}^0)\right]. \tag{2.8}$$

Fluctuation of the state. $P(R)$ is the probability of the given set $\boldsymbol{\rho}$ of relative frequencies in a long series of N measurements. It has the form

$$P(R) = [P(V)]^N \tag{2.9}$$

of the probability that N-times an event V occurs in the series, indepently of one another. In a single observation it results with the probability

$$P(V) = \exp\left[-K(\boldsymbol{\rho}, \boldsymbol{\rho}^0)\right]. \tag{2.10}$$

We interprete V as the "event" that instead of $\boldsymbol{\rho}^0$ another distribution $\boldsymbol{\rho}$ is observed in the form of relative frequencies.

We shall identify a fluctuation, observable only for a short time, in a long lasting thermal equilibrium $\boldsymbol{\rho}^0$ with such an event V. Thus again we are led to the information gain. Now we can refer to the general connection between probability P and bit-number

$$b = -\ln P, \tag{2.11}$$

and obtain again, independently of the introduction of the information gain in chapter 1.2.2, that $K(\boldsymbol{\rho}, \boldsymbol{\rho}^0)$ is the bit-number needed to correct the probability distribution $\boldsymbol{\rho}^0$ into $\boldsymbol{\rho}$.

The corresponding is also true in quantum mechanics where $\boldsymbol{\rho}$ and $\boldsymbol{\rho}^0$ are statistical operators [19]. This was explained in chapter 2.1.4. Therefore we can apply eq. (2.10) to quantum mechanical mixture states as well.

Probability density in the parameter space of fluctuations. We are interested in the probability density in the space of the macroscopic thermal variables M^ν describing a fluctuation. These variables are mean values of phase space functions $\mathbf{M}^\nu$. If we intro-

[19] *F. Schlögl*, Z. Physik **249**, 1 (1971).

duce probabilities of these variables, we change from the level of microstates in Γ-space as events to a new level of random events, the fluctuations. The variables M which were mean values in the microscopic description, are now random quantities. The fluctuations are so-called *mesoscopic* states because they represent a level of description lying between the macrosopic and the microscopic level.

The generalized canonical distributions ρ in the Γ-space corresponding to given thermal variables M represent a relative small subset of the large manifold of all possible distributions in Γ-space. The correspondence between these particular ρ and the set M of the thermal variables M^ν is a one-to-one mapping. We identify the event V, of which the probability is given by eq. (2.10), with a fluctuation described by the set of the parameters

$$V^\nu = M^\nu - M^{0\nu},\tag{2.12}$$

where $M^{0\nu}$ are the values of the thermal variables M^ν if taken for the equilibrium ρ^0. (The notation V^ν has not to be mistaken for the notation V^l we used for the working parameters in previous chapters.) In this concept, ρ is a short-living thermal state of the type of an "inhibited" or "frozen equilibrium", whereas ρ^0 is a long lasting unrestricted equilibrium.

According to the deduction of eq. (2.10), there the events V represented a discrete set, but now the fluctuations V are continuous. We can start with a discrete set by dividing the space of the parameters M into small cells of equal size. The metrics defining the size of the cells in principle can depend on the special kind of the chosen thermal varibales M^ν. We assume that the fluctuations are small enough to neglect a possible difference between the local metrics of macroscopic M-space in the two points M and M^0. This is always the case in a macroscopic system if we exclude phase transitions because then the fluctuations are very small on a macroscopic scale, as the results of the preceding chapter 5.1.1 have shown. The generalized canonical distributions shall only be discriminated if belonging to different probabilities of the cells. For these discrete sets we can apply eq. (2.10). Letting the size of the cells go to zero leads to continuous V and we obtain the probability density of fluctuations in V-space

$$W(V) = C \exp\left[-K(V)\right],\tag{2.13}$$

where $K(V)$ is $K(\rho, \rho^0)$ expressed as a function of the set V. According to eq. (3.10) of chapter 1.2.3 it holds that

$$K(V) = \Xi(\zeta) - \zeta_\nu V^\nu\tag{2.14}$$

with

$$\zeta_\nu = -\frac{\partial K}{\partial V^\nu}.\tag{2.15}$$

Einstein's fluctuation formula. We discuss in particular a microcanonical distribution of an isolated system. Fluctuations will not deviate from the energy shell. Therefore we can make the restriction that all allowed microstates are on the shell, where ρ^0 is constant. In the general expression

$$K(\rho, \rho^0) = I(\rho) - I(\rho^0) - \text{tr}\left[(\rho - \rho^0)\ln\rho^0\right]\tag{2.16}$$

then the last term is zero, and we obtain the *Einstein fluctuation formula*:

$$W(V) = C \exp \frac{\Delta S}{k} \tag{2.17}$$

where

$$\Delta S = S - S^0 \tag{2.18}$$

is the entropy change associated with the formation of the fluctuation. The fluctuation is less probable than the state ρ^0, and ΔS is negative. As in the microcanonical distribution temperature T is constant like energy U, we can also write

$$W(V) = C \exp \left(-\frac{\Delta F}{kT} \right), \tag{2.19}$$

where ΔF is the increase of *Helmholtz* free energy caused by the fluctuation.

Correlations of the conjugates. Let us return to the more general case of eq. (2.13). As a consequence of eqs. (2.14, 2.15) it holds that

$$\frac{\partial W}{\partial V^\nu} = \zeta_\nu W. \tag{2.20}$$

On the mesoscopical level, ζ_ν is also a random quantity like V^ν. We can call V^ν, ζ_ν "thermally conjugate" fluctuation variables. With eq. (2.13) we find the correlation matrix of V and ζ:

$$\langle V^\mu \zeta_\nu \rangle = \int dV\, W\, V^\mu \zeta_\nu = \int dV\, V^\mu \frac{\partial W}{\partial V^\nu}. \tag{2.21}$$

Infinitely large fluctuations V have to be impossible in the sense that W has to vanish sufficiently fast in the infinity of the V-space. Therefore partial integration yields

$$\langle V^\mu \zeta_\nu \rangle = -\delta^\mu_\nu, \tag{2.22}$$

where δ^μ_ν are the *Kronecker* symbols, the elements of the unit matrix. In particular we see that V^μ and ζ_ν are uncorrelated if $\mu \neq \nu$.

5.1.3   Applications of *Einstein*'s Fluctuation Formula

Galvanometer mirror. In special galvanometers a mirror is used which is suspended by a torsion file. Due to the elasticity of the material of the file, the mirror is bound to a rest position by a turning moment proportional to the torsion angle α from the rest position. The increase ΔF of *Helmholtz* free energy induced by an angle α is equal to the mechanical work

$$\Delta F = D \frac{\alpha^2}{2}. \tag{3.1}$$

D is the *torsion constant*. The usually small and light mirror is very sensitive to thermal fluctuations in the surrounding gas caused by irregular impacts of molecules of the gas.

They give rise to fluctuations of the angle α observed by a reflected light beam. The probability density of a fluctuation α according to the *Einstein* formula is

$$W(\alpha) = C \exp\left(-\frac{D\alpha^2}{2kT}\right). \qquad (3.2)$$

This is a normal distribution over α with the mean square

$$\langle \alpha^2 \rangle = \frac{kT}{D}. \qquad (3.3)$$

This result allows us to measure that *Boltzmann* constant k and to yield the *Loschmidt* number

$$L = \frac{R}{k} \qquad (3.4)$$

by means of the gas constant R.

Barometric pressure. The barometric pressure formula has already been derived in different ways in earlier parts of this book. We also can obtain it as a consequence of the *Einstein* formula. We consider an arbitrary particle in gaseous or liquid surroundings in thermal equilibrium. To raise the particle reversibly from an equilibrium position to the height z in the gravitational field requires the work

$$\Delta F = mgz. \qquad (3.5)$$

m is the particle mass, and g the gravitational acceleration. *Einstein*'s formula eq. (2.19) yields

$$W(z) = C \exp\left(-\frac{mgz}{kT}\right). \qquad (3.6)$$

The ratio of probabilities of two positions z_1, z_2 is

$$\frac{W(z_2)}{W(z_1)} = \exp\left[-\frac{mg}{kT}(z_2 - z_1)\right]. \qquad (3.7)$$

This equation gives not only the ratio of the density of atmospheric air in different heights but can be applied also to the density of the number of macroscopic particles, as for instance mastix droplets in water. These can be seen and counted by means of an optical microscope. In this case, m is so large that remarkable changes of density occur over some 10 cm leading us to a further method of measuring the *Boltzmann* constant.

Formation of droplets. In chapter 2.3.7, the dependence of vapor pressure over droplets upon the radius was discussed. A particular result was that droplets in a vapor of a given pressure p will increase in size only if their radius is larger than a certain "critical" radius r_0. Such droplets then are nuclei of condensation. On account of eq. (7.31) of chapter 2.3.7, this radius is determined by

$$\ln \frac{p}{P_\infty} = \frac{2\sigma v'}{kTr_0}. \qquad (3.8)$$

In this context we use the notation v' for the volume of the fluid per molecule, not per mole. Let us now consider the probability that a spherical droplet of the critical size is formed by thermal fluctuations. To apply the *Einstein* formula in the form

$$W(r_0) = C \exp\left(-\frac{\Delta F}{kT}\right), \tag{3.9}$$

we need the change ΔF of *Helmholtz* free energy necessary for the formation of the droplet out of the vapor. The droplet may contain n molecules. The formation can be accomplished as a gedanken experiment in the following four steps of a reversible isothermal process. First, removal of n molecules from the vapor at pressure p isothermally by means of a piston. Second, expansion to pressure P_∞. Third, condensation over a plain surface at pressure P_∞. Fourth, formation of the droplet out of the liquid, say by means of a pipette. We introduce some approximations. We assume that the vapor is an ideal gas. Then the work nkT of step one is cancelled by step three. Moreover, we neglect the difference between the mole volume of the liquid at the different pressures p and P_∞. Then the formation of the droplet in step four requires only the work $\sigma\omega$ if ω is the surface and σ the surface tension of the droplet. The ideal gas assumption also enters into the expansion work of step two and we obtain

$$\Delta F = -nkT \ln\frac{p}{P_\infty} + \sigma\omega. \tag{3.10}$$

If the droplet has the critical size, the external pressure p is equal to the corresponding vapor pressure and the following equation holds

$$nkT \ln\frac{p}{P_\infty} = \frac{2\sigma}{r_0} nv' = \frac{2\sigma}{r_0} \frac{4\pi}{3} r_0^3 = \frac{2}{3}\sigma\omega, \tag{3.11}$$

yielding

$$\Delta F = \frac{1}{3}\sigma\omega = \sigma\frac{4\pi}{3} r_0^2. \tag{3.12}$$

With the abbreviation

$$a^2 = \frac{3kT}{4\pi\sigma} \tag{3.13}$$

we obtain

$$W(r_0) = C \exp\left[-\left(\frac{r_0}{a}\right)^2\right]. \tag{3.14}$$

The dimensionless number

$$x = \ln\frac{p}{P_\infty} \tag{3.15}$$

is defined to be the *supersaturation* of the vapor and is a common quantity in meteorology. It is connected with r_0 by

$$x = \frac{2\sigma v'}{kTr_0}. \tag{3.16}$$

With

$$A = \left(\frac{2\sigma v'}{kTa}\right)^2 \tag{3.17}$$

we may write

$$W(r_0) = C \exp\left(-\frac{A}{x^2}\right). \tag{3.18}$$

For water at 275 °K, we obtain $A = 115$. This means that already a small change δx of the supersaturation x gives rise to a very large change of

$$\frac{\delta W}{W} = \frac{2A}{x^2} \frac{\delta x}{x}. \tag{3.19}$$

Therefore it does not matter whether we considered only the probability of a critical droplet and not that of any larger one.

The theory of the formation of condensation nuclei is called "nucleation theory". Our results, however, are based on several idealizations. One is that we assumed only spherical droplets. Also if this is realistic, the results will be applicable only to extremely pure vapor. Any light pollution gives rise to another formation of nuclei which is predominant compared to the thermal nucleation considered here. A typical example of a nonthermal production of droplets is the nucleation by ionized molecules along the path of a high energy particle in a cloud chamber. Another example is the nucleation in the jet trail of an airplane.

*5.1.4 *Maxwell*'s Demon

In 1871, *J. Clerk Maxwell* raised a widely discussed question. It is connected with a fictitious being, today called *Maxwell's demon*. This demon is supposed to be able to observe the single molecules in a gas and to separate them in the following way. He can open and close an opening in a wall which separates two parts of a vessel filled with the gas and can decide whether to let pass an arriving molecule or not. We consider the following process. If a molecule came from one side, the demon would open the opening, if it came from the other side, he would close the opening. This procedure would concentrate the gas in one part and rarefy it in the other part. The question now is, if the existence of such a being would contradict the Second Law.

In 1929 *L. Szilard* gave a conclusive answer which for the first time revealed a connection between information and thermodynamic entropy, two decades before the development of information as a quantitative concept[20].

We represent *Szilard*'s consideration in a changed version. To register an observation, whichever physical system has to be brought into an excited state. Today, in the time of computers, we should say, a bit has to be stored in a storage unit. The excitation energy has to be distinctly higher than the energy the thermal fluctuations have in

---

[20] *L. Szilard*, Z. Physik **53**, 840 (1929).
Compare also: *L. Brillouin*, Vie, matière et observation (Albin Michel, Paris 1959);
W. Büchel, Physikalische Blätter **18**, 400 (1962).

the average. Otherwise the storage would not survive the time necessary for the registration. This means that the probability of this excited state occurring as a thermal fluctuation has to be considerably smaller than one, say smaller than a certain value $1/C$. If ΔS_1 is the change of entropy associated with the excitation, we have to require

$$\exp \frac{\Delta S_1}{k} \leqslant \frac{1}{C} \, . \tag{4.1}$$

It should be mentioned that ΔS_1 is negative. The excitation needs a work W done on the system for which the relation

$$W \geqslant - T \, \Delta S_1 \tag{4.2}$$

holds. After a certain time, the registering system has to be brought into the initial state again to be ready for the next registration. This means that the storage unit has to be made empty. By this process, however, W is transformed into energy in a form which is not anymore arbitrarily available for work, and an entropy increase

$$\Delta S \geqslant \frac{W}{T} \geqslant k \ln C \tag{4.3}$$

is generated anywhere.

Expressed in the terminology of information theory, developed two decades after *Szilard*'s paper, this entropy increase ΔS is the price we have to pay for one bit. For B bits this is an entropy increase

$$\Delta S \geqslant B k \ln C = k \, \Delta I \frac{\ln C}{\ln 2} \, . \tag{4.4}$$

If we choose C equal to 2, we obtain the conventional (but arbitrary) definition of the unit of *Shannon* information for ΔI.

We see that the existence of *Maxwell*'s demon would not contradict the Second Law. This is the result of the consideration which was stressed by *Szilard*. Today, however, we consider the connection between entropy and information to be a result which is no less important, in particular as it was found by *Szilard* before the development of communication theory.

5.2   Nonequilibrium Dynamics

In order to find an adequate description of thermal nonequilibrium states by probability distributions in phase space, we shall discuss the full microscopic dynamics of an isolated system on the most detailed level. This will be done in the first chapter of this subsection. The most detailed level of description is the *Hamilton* dynamics in the Γ-space of classical mechanics or the dynamics in the *Hilbert* space of quantum mechanics. These two cases are often uniformily called *Liouville dynamics* with regard to the *Liouville* theorem which we discussed in chapter 2.1.2. We shall develop a uniform description for these two cases by use of the so-called *Liouville operator* and draw some consequences from its properties. In the second chapter, we shall use this operator for the construction of a probability distribution to a nonequilibrium state, the so called

Mori distribution, which represents an adequate description for a system with a distinct time scale separation. The use of the *Mori* distribution will lead us to a justification of the local equilibrium approximation and show its limits. In the third of the following chapters, we shall discuss an example of quantum dynamics of a nonisolated system which is in contact with a heat reservoir. The system will be described by the quantum mechanical *Wangsness-Bloch* equation. Because we then shall consider a two-dimensional state space only, this description will become relatively simple. The last chapter of this subsection is devoted to an uncertainty relation between thermally conjugate quantities which can be generalized to nonequilibrium distributions.

*5.2.1   The *Liouville* Operator

The *Heisenberg* picture. We consider a conservative mechanical system with the *Hamiltonian* $H(q,p)$. The dynamics of this system is determined by the canonical equations

$$\dot{p}_k = -\frac{\partial H}{\partial q_k}, \qquad \dot{q}_k = \frac{\partial H}{\partial p_k}, \tag{1.1}$$

which we already used in chapter 2.1.2. With the introduction of the so-called *Liouville operator*

$$\mathscr{L} = \frac{1}{i} \sum_k \left(\frac{\partial H}{\partial p_k} \frac{\partial}{\partial q_k} - \frac{\partial H}{\partial q_k} \frac{\partial}{\partial p_k} \right) \tag{1.2}$$

we can write for any observable $M(q,p)$

$$\frac{\mathrm{d}M}{\mathrm{d}t} = i\,\mathscr{L}M. \tag{1.3}$$

(i is the imaginary unit.) The *Liouville* operator $\mathscr{L}$ is a linear operator in the space of the so-called "phase functions" $f(\xi)$ of the phase space variables

$$\xi = (\xi_1 \dots \xi_{2f}) = (q_1 \dots q_f, p_1 \dots p_f). \tag{1.4}$$

The factor $1/i$ in the definition is convenient for the following reasons.

It is advantageous to define a scalar product of two phase functions f, g by

$$[f,g] = \int \mathrm{d}\xi\, f^* g \tag{1.5}$$

and to restrict the space of the used phase functions to those for which these integrals always exist. This involves that the functions are normalizable, which means that the scalar product of a function with itself is finite. A one-to-one mapping of these functions to a vector space is possible, which in general has infinite dimensionality. We say the function space "is" a vector space, and we can adopt all relations which follow from the structure of a vector space. By the definition eq. (1.5) of the scalar product, the metrics of the space is *Hermitian*. This metrics is familiar to physicists from quantum mechanics and allows us to apply a convenient formalism. The elements of the "vector space" are functions of the dynamical variables which themselves form the Γ-space. The Γ-space is of infinite extension. In particular the values of the momenta p_k are not restricted. The integral in eq. (1.5) is extended over the infinite Γ-space and has to be convergent. This requirement that the integral in the scalar products exists, involves

$$\int d\xi \, \mathscr{L}(fg) = 0, \tag{1.6}$$

due to the way in which the differentiations appear in $\mathscr{L}$. Moreover,

$$\mathscr{L}(fg) = f\mathscr{L}g + g\mathscr{L}f, \tag{1.7}$$

and we obtain

$$[f, \mathscr{L}g] = - \int d\xi \, g \, \mathscr{L}f^* = - (\mathscr{L}^*f, g). \tag{1.8}$$

Generally, $\mathscr{V}^+$ is called the *Hermitian adjoint* operator to a linear operator $\mathscr{V}$ if the following relation holds for all f, g:

$$[f, \mathscr{V}g] = [\mathscr{V}^+f, g]. \tag{1.9}$$

Due to the inclusion of the factor $1/i$ into the definition eq. (1.2) of the *Liouville* operator, the complex conjugate $\mathscr{L}^*$ of $\mathscr{L}$ is equal to $-\mathscr{L}$ and we obtain

$$[f, \mathscr{L}g] = [\mathscr{L}f, g], \tag{1.10}$$
$$\mathscr{L}^+ = \mathscr{L}. \tag{1.11}$$

This means that $\mathscr{L}$ is a *Hermitian* operator.

A probability distribution in phase space which is based on observations, say at a time zero, is represented by a probability density $\rho(\xi)$ in Γ-space. The mean value of an observable $M(\xi)$ is

$$\langle M \rangle = \int d\xi \, \rho M. \tag{1.12}$$

Its derivative with respect to time is

$$\frac{d}{dt}\langle M \rangle = \left[\rho, \frac{dM}{dt}\right] = [\rho, i\mathscr{L}M]. \tag{1.13}$$

We have to stress that in this description we assume that an observable M is described by the same phase space function $M(\xi)$ at any time if defined by the same measuring prescription. In the following we speak only of such observables and do not include an explicit dependence on t. Such observables M change with time only implicity because the mapping point ξ of the system moves in the phase space. (An explicit dependence would be possible if the measuring prescription changes or if an external influence is present, like for instance a time dependent magnetic field. Then, however, the system were not conservative any more.)

We have to stress that this scheme is not the only one to describe the time dependence of the system. By this specific formalism in which observables M change with time, yet ρ is time independent, the so-called *Heisenberg picture* is defined. This name is borrowed from quantum mechanics, as is the name of the following picture, too, which is another possible scheme.

The *Schrödinger* picture. Physical meaning has only the time dependence of mean values. Therefore there is a certain arbitrariness of how the time dependence is shared among state ρ and observable M in the formal description. The time derivative of a mean value $\langle M \rangle$ has to be the same. We can write it in form of eq. (1.13) or in the form

$$\frac{\mathrm{d}}{\mathrm{d}t} \langle M \rangle = -\left[i\mathscr{L}\rho, M\right] \tag{1.14}$$

as well. This is a consequence of eq. (1.10). It is now possible to describe the time dependence alternatively by

$$\frac{\mathrm{d}\rho}{\mathrm{d}t} = -i\mathscr{L}\rho, \tag{1.15}$$

$$\frac{\mathrm{d}M}{\mathrm{d}t} = 0. \tag{1.16}$$

This means that we have shifted the time dependence from M to ρ. It corresponds to the alternative view that observables M defined by the same measuring process at any time should be described by a time independent phase function. This formalism is called the *Schrödinger picture*.

We were speaking of conservative systems which have a time independent *Hamiltonian*. For such systems the *Heisenberg* and the *Schrödinger* picture both are possible. If we, however, envisage the extension to open systems, in general only the *Schrödinger* picture will be adequate. Then external influences will lead us to a change with time of the knowledge contained in ρ, also after the last observation of the system itself. Then ρ becomes necessarily time dependent and in general it will no more be possible to shift the time dependence of the mean values of observables into a time dependence of the observables only by a comparably simple transformation.

Eigenfunctions of the *Liouville* operator. As the *Liouville* operator $\mathscr{L}$ is a *Hermitian* operator in our space of phase functions, there exists a system of orthogonal normalized eigenfunctions $\psi_\nu(\xi)$ with real eigenvalues l_ν satisfying the equation

$$\mathscr{L}\psi_\nu(\xi) = l_\nu\psi_\nu(\xi). \tag{1.17}$$

It should be stressed that in general these ψ_ν are not positive probability distributions. They can even assume complex values. Any probability distribution ρ can be represented as a linear combination of the eigenfunctions ψ_ν:

$$\rho = \sum_\nu c_\nu \psi_\nu. \tag{1.18}$$

In the *Schrödinger* picture the coefficients c_ν are time dependent, the eigenfunctions ψ_ν are not. Eq. (1.15) yields

$$\frac{\mathrm{d}}{\mathrm{d}t} c_\nu = -i l_\nu c_\nu, \tag{1.19}$$

$$c_\nu(t) = c_\nu(0)\exp\left(-il_\nu t\right), \tag{1.20}$$

$$\rho(t) = \sum_\nu c_\nu(0)\,\psi_\nu\exp\left(-il_\nu t\right). \tag{1.21}$$

We call to mind the definition of a function of a *Hermitian* operator in quantum mechanics by eq. (4.27) of chapter 2.1.4. This definition is valid in any *Hermitian* vector space and can also be applied to the classical *Liouville* operator. It yields in particular

$$\exp\left(-i\mathscr{L}t\right)\psi_\nu = \psi_\nu \exp\left(-il_\nu t\right). \tag{1.22}$$

Therefore we can write

$$\rho(t) = \exp\left(-i\mathscr{L}t\right)\rho(0). \tag{1.23}$$

We obtain directly with eq. (1.19):

$$\frac{d\rho}{dt} = -i\mathscr{L}\rho. \tag{1.24}$$

In the *Heisenberg* picture instead of this equation, the following holds

$$\frac{dM}{dt} = i\mathscr{L}M. \tag{1.25}$$

Liouville operator in quantum mechanics. In quantum mechanics, the *Schrödinger* equation of a system with the *Hamilton* operator **H** has the form

$$\frac{d}{dt}\,|\alpha\rangle = -\frac{i}{\hbar}\,\mathbf{H}\,|\alpha\rangle \tag{1.26}$$

for a pure state $|\alpha\rangle$ and defines the *Schrödinger* picture ($\hbar$ is *Planck*'s constant h divided by 2π). The *Schrödinger* equation yields for a statistical operator

$$\rho = \sum_\alpha |\alpha\rangle P_\alpha \langle\alpha| \tag{1.27}$$

the dynamical equation

$$\frac{d\rho}{dt} = -\frac{i}{\hbar}\,(\mathbf{H}\rho - \rho\mathbf{H}) = -i\mathscr{L}\rho. \tag{1.28}$$

This has the same form as eq. (1.24) with introduction of the quantum mechanical *Liouville* operator $\mathscr{L}$ through

$$\mathscr{L}\mathbf{A} = \frac{1}{\hbar}\,(\mathbf{HA} - \mathbf{AH}) \tag{1.29}$$

with an arbitrary linear operator **A** in *Hilbert* space of the wave functions $|\alpha\rangle$. We have to make the following distinction: the operators in *Hilbert* space like ρ, like the observables **M**, or like any operator **A** act on wave functions $|\alpha\rangle$. The *Liouville* operator, however, acts on such linear *Hilbert* space operators **A** and is sometimes called a *superoperator*. It is in particular a linear superoperator. Generally, to describe such a linear superoperator $\mathscr{V}$ in detail, we consider the matrix elements $A_{\alpha\beta}$ of **A** in an orthogonal basis system $|\alpha\rangle$. For these,

$$(\mathscr{V}\mathbf{A})_{\alpha\beta} = \sum_{\gamma\delta} V_{\alpha\beta\gamma\delta}\,A_{\gamma\delta} \tag{1.30}$$

holds. Thus a linear superoperator is represented in a basis system by a tensor of order four; in particular the *Liouville* operator by

$$L_{\alpha\beta\gamma\delta} = \frac{1}{\hbar}\,(H_{\alpha\gamma}\delta_{\beta\delta} - \delta_{\alpha\gamma}H_{\delta\beta}). \tag{1.31}$$

This also gives the key to performing the integration of eq. (1.28) in detail, which yields in accordance with eq. (1.23) the equation

$$\rho(t) = \exp(-i\mathscr{L}t)\,\rho(0), \tag{1.32}$$

often called the *J. von Neumann equation*. For an expectation value of an observable **M**, we obtain

$$\frac{\mathrm{d}}{\mathrm{d}t}\,\langle M\rangle = \frac{\mathrm{d}}{\mathrm{d}t}\,\mathrm{tr}(\mathbf{M}\rho) = -i\,\mathrm{tr}(\mathbf{M}\mathscr{L}\rho). \tag{1.33}$$

The change to the *Heisenberg* picture is performed easily by the relation

$$\mathrm{tr}(\mathbf{M}\mathscr{L}\rho) = -\mathrm{tr}(\rho\mathscr{L}\mathbf{M}) \tag{1.34}$$

and eq. (1.29). In the *Heisenberg* picture the following equations hold:

$$\frac{\mathrm{d}\mathbf{M}}{\mathrm{d}t} = i\mathscr{L}\mathbf{M}, \tag{1.35}$$

$$\frac{\mathrm{d}\rho}{\mathrm{d}t} = 0, \tag{1.36}$$

$$\mathbf{M}(t) = \exp(i\mathscr{L}t)\mathbf{M}(0). \tag{1.37}$$

In the eigenbasis of the energy operator, the *Hamiltonian* is diagonal

$$H_{\alpha\beta} = E_\alpha\delta_{\alpha\beta}, \tag{1.38}$$

and the superoperator occurring in eq. (1.37) takes the form

$$[\exp(i\mathscr{L}t)]_{\alpha\beta\gamma\delta} = \delta_{\alpha\gamma}\delta_{\beta\delta}\exp\left[\frac{i}{\hbar}(E_\alpha - E_\beta)t\right]. \tag{1.39}$$

The equations (1.35, 1.36, 1.37) for the *Heisenberg* picture and the equations (1.23, 1.24, 1.25) for the *Schrödinger* picture are the same in classical as in quantum physics. This makes the introduction of the *Liouville* operator in the following very useful for the common description.

Invariance of entropy. The following holds in classical as well as in quantum physics. In the *Heisenberg* picture, where ρ is time independent, it is a trivial statement that entropy

$$S = -k\,\mathrm{tr}(\rho\ln\rho) \tag{1.40}$$

is invariant with respect to *Hamilton* dynamics, which with regard to classical physics is often called *Liouville dynamics*.

Remark: Not only S but all cumulants C_k of the bit-number

$$b = - \ln \rho \tag{1.41}$$

of any order are *Liouville invariants*. This follows directly from the invariance of the generating function

$$\Gamma(\alpha) = \ln \langle \exp(\alpha \mathbf{b}) \rangle = \ln \operatorname{tr} \rho^{1-\alpha} = \sum_{k=1}^{\infty} \frac{\alpha^k}{k!} C_k \tag{1.42}$$

of the cumulants. If any one of these cumulants changes with time, this change is an indication of a deviation from the *Liouville* dynamics.

5.2.2   The *Mori* Distribution

We consider a thermal nonequilibrium state which is described in the macroscopic scope by local time dependent variables $M^\nu(\vec{x}, t)$. To give an example, these variables may be associated with a heat conductive process. Then M^ν is the density of energy in ordinary space at space point $\vec{x}$ and time t. Another example is a diffusion process for which M^ν is the local concentration of the diffusing substance in a solution. In the statistical theory we interpret these densities $M^\nu(\vec{x}, t)$ as statistical mean values. In the following we shall be confronted with the problem of finding the probability distribution in Γ-space which describes the instantaneous thermal state in an adequate way.

First the question arises whether the *Jaynes* method of unbiased guess is appropriate if based on the knowledge only of the instantaneous local mean values M^ν. It yields

$$\rho_L(t) = \exp \int d^3 x \left[\Psi(\vec{x}, t) - \lambda_\nu(\vec{x}, t) \mathbf{M}^\nu(\vec{x}) \right]. \tag{2.1}$$

In this expression, $\mathbf{M}^\nu$ are phase space functions in classical physics and operators in *Hilbert* space in quantum physics. They are densities in ordinary $\vec{x}$-space. This distribution ρ_L is called a *local equilibrium distribution* because it describes the system to be in a *local equilibrium* in any sufficiently small cell of ordinary space at time t. In this thermal equilibrium, exclusively in the cell, the variables $\lambda_\nu(\vec{x}, t)$ are the thermal conjugate to the densities $M^\nu(\vec{x} t)$. In more general cases, the name *accompanying equilibrium* for ρ_L of eq. (2.1) would be even more adequate if, like for the following, the fundamental property is the dependence on time yet not on space.

This distribution, however, is not satisfactory. It is based on the instantaneous mean values M^ν only and does not take into account any further knowledge about the dynamical properties of the system. Therefore it is not able, for instance, to give an adequate description of time correlations in the system. In particular it fails totally to describe transport processes correctly.

In the following we consider systems with a distinct time scale separation. As already explained, we suppose that in the system of macroscopic dimensions, the microscopic processes are very fast compared with the macroscopically observed processes. Microscopic processes for instance are collisions of molecules, chemical reactions between single molecules, or also common oscillations of only few molecules. We can practically describe microscopic processes solely if only few degrees of freedom are involved, whereas the macroscopically describable processes are connected with collective motions of large numbers of molecules. Such a macroscopic process is for instance the drift of a fluid, the deformation of a solid, or the chemical conversion in a reactor. The distinct time scale separation means that there is a long scope of time intervals which are very short compared with the macroscopic processes and very long compared with the microscopic processes. Often it is said that these intermediate times form a *plateau*. The time scale separation, i.e. the existence of such a plateau, is also called *Onsager separation*. There are macroscopic systems which never have this property. One example is the radiation cavity. The macroscopically observable propagation of light does not occur slower than the exchange of energy between different radiation modes; quite the contrary. Another example is the *Knudsen* gas which is so highly rarefied that collisions between molecules happen no more frequently than collisions with the walls of the vessel. The exclusion of such counter examples, i.e. the assumption of the *Onsager* separation, is a fundamental condition for the possibility of describing nonequilibrium processes by local thermal variables. Notwidthstanding that this property is assumed to be fulfilled as a rule for material systems, it is not proved as a consequence of microdynamics. There are only very few model systems for which it is possible to show that this property is connected with the large number of degrees of freedom. A standard model will be demonstrated in chapter 5.4.

In 1958 the Japanese physicist *H. Mori* proposed a construction of a time dependent phase distribution for systems with the required time scale separation[21]. τ is a time lying on the plateau if

$$t_{mi} \ll \tau \ll t, \tag{2.2}$$

where t_{mi} is a time on the microscopic short time scale, and t is a macroscopically observable time. During the time τ practically no remarkable influence of the environment on the system becomes active, affecting at most a negligible thin surface layer of the system. Therefore the system is practically isolated during this time and its dynamics is ruled by a *Hamiltonian* and thus by a *Liouville* operator $\mathscr{L}$. We take the local equilibrium distribution ρ_L at a time $t - \tau$ and apply to this the reversible microdynamics during the time τ. This can be expressed with the help of the *Liouville* operator. The result is the *Mori distribution*

$$\rho(t) = \exp(-i\mathscr{L}\tau)\rho_L(t - \tau). \tag{2.3}$$

The construction is illustrated schematically in Figure 50. $M(x)$ may be for instance the energy density on a heat conducting bar extended in x-direction. It may have the shape (b) at time t. The local equilibrium at time $t - \tau$ corresponds to (a) in which each cell of the bar is isolated and is in an equilibrium of its own. If the isolating walls between the cells are removed, after a time τ the *Liouville* dynamics will have developed the full internal flows, yet will not have changed the macroscopic shape $M(x)$. Later on, this shape will change only after a time on the long time scale into say (c). The result will be independent of the special choice of τ, provided this lies on the plateau.

---

[21] *H. Mori*, Phys. Rev. **115**, 298 (1959).

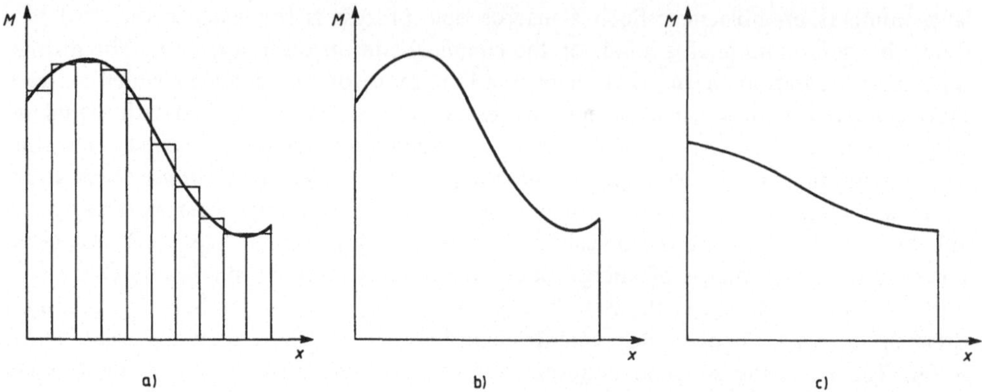

Figure 50   Schematical representation of the *Mori* construction of the probability distribution in phase space corresponding to a thermal nonequilibrium state. Starting with a local equilibrium (a) in which isolated cells are in their individual equilibria the application of the full *Liouville* dynamics yields a distribution (b) in which after a time τ all interior flows could develop to the *Mori* distribution (b) with unchanged macroscopic variables. These will change into (c) in a macroscopic time.

The *Mori* construction has been worked out in more details and was successfully used for the calculation of transport coefficients. We shall use it only to show very general results for which eq. (2.3) suffices. As the thermal variables M^ν practically do not change during the time τ, they are the same for $\rho(t)$ as for $\rho_L(t-\tau)$ and moreover also the same as for $\rho_L(t)$. This is again true for the entropy which remains unchanged by the reversible motion, the *Liouville* motion. By the way, this is true not only for entropy but for all bit-number cumulants C_k of eq. (1.12), in agreement with the remark at the end of chapter 5.2.1. It is not true, for instance, for quantities which are essentially dependent on correlations in the system.

So we obtain the result that entropy of the *Mori* distribution has the same functional dependence on the thermal variables as entropy of the local equilibrium distribution. As the M^ν are local densities, we can define a local entropy density $S(\vec{x}, t)$ which satisfies the relation

$$\frac{1}{k} S = \lambda_\nu M^\nu - \Psi. \tag{2.4}$$

Unlike the time independent operators $\mathbf{M}^\nu$ in the *Schrödinger* picture of eq. (2.1), the M^ν occurring here are time dependent mean values. The general relations for the generalized canonical distribution are now valid at any point $\vec{x}$ in space and at any time t:

$$\frac{\partial S}{\partial M^\nu} = k\lambda_\nu, \tag{2.5}$$

$$\frac{\partial \Psi}{\partial \lambda_\nu} = M^\nu. \tag{2.6}$$

Eq. (2.5) is the local *Gibbs* fundamental equation.

*5.2.3   The *Wangsness-Bloch* Equation

In this chapter we shall consider a quantum mechanical equation which can be used to describe, in a phenomenological approximation, the dynamics of a nonisolated system in a time independent thermal environment. We assume that the influence of the environment causes a regression to its thermal equilibrium. Therefore an ansatz of the following form is made for the dynamical equation of the statistical operator ρ of the system in the *Schrödinger* picture

$$\frac{d\rho}{dt} = -i\,\mathcal{L}\rho + \mathcal{D}\,(\rho - \rho^0). \tag{3.1}$$

In macroscopic thermodynamics, the use of these equations is called the *local equilibrium approximation*. Whereas in statistics the use of the local equilibrium distribution ρ_L is deficient for other relations, the use of the equations (2.4, 2.5, 2.6) is justified whenever the far better motivated use of the *Mori* distribution is adequate.

Another important quantity which is equal for the *Mori* and the local equilibrium distribution is the information gain $K(\rho, \rho')$ of two mixture states of the same system because they are subject to the same *Liouville* operator $\mathcal{L}$ in eq. (2.3). Thus also the use of the local availability obtained with the local equilibrium is justified in the same way as the use of the local entropy.

$\mathcal{L}$ is the quantum mechanical *Liouville* operator of the system itself, defined by

$$\mathcal{L}\rho = \frac{1}{\hbar}\,(\mathbf{H}\rho - \rho\mathbf{H}). \tag{3.2}$$

$\mathbf{H}$ is the *Hamiltonian* of the system if not in contact with the environment. ρ^0 is the state of the system if in equilibrium with the environment, and $\mathcal{D}$ is a time independent linear superoperator, as $\mathcal{L}$ is, too. As ρ always has to be normalized, the relation

$$\mathrm{tr}\,\frac{d\rho}{dt} = 0 \tag{3.3}$$

holds and we have to require that

$$\mathrm{tr}\,\mathcal{D}\,(\rho - \rho^0) = 0. \tag{3.4}$$

Moreover, $\mathcal{D}\,(\rho - \rho^0)$ has to be an *Hermitian* operator in *Hilbert* space. Under these restrictions, eq. (3.1) is called the *Wangsness-Block equation*[22].

We shall apply eq. (3.1) to the spin of an atomic nucleus in surrounding matter, and under the influence of an external magnetic field B in x_3-direction. The nuclear spin in particular shall be 1/2. Then the quantum mechanical state space of the spin is two-dimensional and all operators in this space can be represented by a linear combination of the unit operator 1 and the three *Pauli* spin matrices σ_1, σ_2, σ_3. These are defined by the algebraic relations

$$\sigma_1 \sigma_2 = -\,\sigma_2 \sigma_1 = i\,\sigma_3, \tag{3.5}$$

$$\sigma_1^2 = 1, \tag{3.6}$$

---

[22] *R. K. Wangsness* and *F. Bloch*, Phys. Rev. **89**, 728 (1953).

and all equations which arise from these relations by the cyclic exchange of the subscripts. The *Hamiltonian* of the free spin is

$$\mathbf{H} = - g \hbar\, \sigma_3 B.$$
(3.7)

As no other space direction besides x_3 is distinguished, we suppose that only 1 and σ_3 occur in $\mathscr{D}$. Higher powers of σ_3 reduce to 1 or σ_3. Thus only ρ, $\sigma_3\rho$, $\rho\sigma_3$, $\sigma_3\rho\sigma_3$ can occur as independent spin operators in $\mathscr{D}\rho$. With constant numbers λ, a, b we obtain:

$$\mathscr{D}(\rho - \rho^0) = -\lambda(\rho - \rho^0) - ia(\sigma_3\rho - \rho\sigma_3) - b(\rho - \sigma_3\rho\sigma_3).$$
(3.8)

ρ_0 commutes with σ_3. Therefore it appears only in the first term on the right hand side. The second term with the multiplier a is proportional to $\mathscr{L}\rho$ and only causes a change of g into a new coefficient g' with which

$$a = (g' - g)B.$$
(3.9)

So we obtain

$$\frac{d\rho}{dt} = i\,\omega(\sigma_3\rho - \rho\sigma_3) - \lambda(\rho - \rho^0) - b(\rho - \sigma_3\rho\sigma_3)$$
(3.10)

with

$$\omega = g'B.$$
(3.11)

Next we ask for the time dependence of the mean value

$$\vec{s} = \mathrm{tr}\,(\vec{\sigma}\rho).$$
(3.12)

We keep in mind the general rule

$$\mathrm{tr}\,(\mathbf{uv}) = \mathrm{tr}\,(\mathbf{vu})$$
(3.13)

for all linear operators $\mathbf{u}$, $\mathbf{v}$ and obtain

$$\mathrm{tr}\,(\sigma_1\sigma_3\rho) = -\mathrm{tr}\,(\sigma_1\rho\sigma_3) = -is_2,$$
(3.14)

$$\mathrm{tr}\,(\sigma_1\sigma_3\rho\sigma_3) = i\,\mathrm{tr}\,(\sigma_2\rho\sigma_3) = -s_1$$
(3.15)

with the corresponding cyclic complements. This yields

$$\frac{ds_1}{dt} = +\omega s_2 - (\lambda + 2b)s_1,$$
(3.16)

$$\frac{ds_2}{dt} = -\omega s_1 - (\lambda + 2b)s_2,$$
(3.17)

$$\frac{ds_3}{dt} = -\lambda(s_3 - s_3^0).$$
(3.18)

We can write this result in the form

$$\frac{ds_1}{dt} = g'\,(\vec{s} \times \vec{B})_1 - \kappa s_1,$$
(3.19)

$$\frac{ds_2}{dt} = g'\,(\vec{s} \times \vec{B})_2 - \kappa s_2,$$
(3.20)

$$\frac{ds_3}{dt} = -\gamma(s_3 - s_3^0).$$
(3.21)

We see the following. The component s_3 in the field direction regresses with the so-called *longitudinal* spin-lattice relaxation time $1/\gamma$. The components perpendicular to the field perform a *precession* around this direction and regress with the *transversal* spin-lattice relaxation time $1/\kappa$ which empirically is remarkably shorter than the longitudinal one. This corresponds to a large positive value b.

*5.2.4 Thermodynamic Uncertainty Relation

Already about 1932, without a quantitative discussion, *N. Bohr*[23] expressed the conjecture that there has to be an uncertainty relation between energy and temperature in thermodynamics. Any measurement of temperature requires an energy exchange between the system and the thermometer. Therefore it makes energy uncertain to some extent. On the other hand, a sharp value of energy of a microstate can occur in different canonical distributions and thus can be associated with different values of temperature. This means that temperature is unsharp if energy is sharp. This relation between energy and temperature is in some respect an analogue to the uncertainty relation between complementary quantities in quantum mechanics.

In the following we shall see that there is an uncertainty relation for any pair of thermally conjugate variables, not only for generalized canonical distributions and equilibria at all[24]. This relation already holds in general statistics, not restricted to physics, as represented in the first section of this book. Nevertheless we shall discuss it only here in context with fluctuations in thermodynamics.

Let us return to eq. (2.10) of chapter 5.1.2. The event V of finding relative frequencies ρ in a given mixture state ρ^0 has the probability

$$P(V) = \exp\left[-K(\rho, \rho^0)\right]. \tag{4.1}$$

A particular consequence of this relation is the following. If ρ, ρ^0 are canonical distributions with temperatures T and T^0 respectively, there is a certain probability that for a short time the thermometer registers the temperature T also if the system is in the state corresponding to the temperature T^0. In this sense we may speak of "temperature fluctuations".

In the same sense we can speak of fluctuations of general intensive quantities. ρ, ρ^0 shall be generalized canonical distributions. In accordance with eq. (2.13) of chapter 5.1.2 we write for the probability density in V-space

$$W(V) = C \exp\left[-K(V)\right], \tag{4.2}$$

where V^ν is the mesoscopic deviation of the mean value M^ν in ρ from the equilibrium value in ρ^0. The corresponding deviation of the thermally conjugate intensity

$$\zeta_\nu = \lambda_\nu - \lambda_\nu^0 \tag{4.3}$$

---

[23] Discussion in *W. Heisenberg*, Der Teil und das Ganze, chapter 9 (R. Piper Verlag, München 1969). Compare also: *J. Lindhard*, The Lessons of Quantum Theory, ed. by *J. de Boer, E. Dahl*, and *O. Ulfbeck* (Elsevier Sci. Publ., Amsterdam 1966).

[24] *F. Schlögl*, J. of Physics and Chemistry of Solids **49**, 679 (1988).

is a random quantity as well on the mesoscopic level. In chapter 5.1.2 we obtained (eq. (2.22))

$$\langle V^\mu \zeta_\nu \rangle = - \delta^\mu_\nu,$$

(4.4)

where the bracket designates the mean value formed with the distribution $W(V)$.

The uncertainty relation is not restricted to equilibrium distributions but holds also in a nonequilibrium *Mori* distribution at any time point. This is a consequence of the equivalence of $K(V)$ in the *Mori* distribution and in the local equilibrium. Therefore we shall discuss the generalization to arbitrary distributions $W(V)$ of fluctuations V. In the generalized case we define a quantity ζ_ν as "conjugate" to V^ν by the equation

$$\frac{\partial W}{\partial V^\nu} = \zeta_\nu W,$$

(4.5)

which occurred already as eq. (2.20) of chapter 5.1.2 in the special case that ρ^0 is an equilibrium. Like V^ν also ζ_ν is a random quantity on the mesoscopic level and satisfies eq. (4.5).

The *Schwarz* inequality. For any real parameter α the function

$$f(\alpha) = \langle (V^\mu + \alpha \zeta_\nu)^2 \rangle = \alpha^2 \langle \zeta_\nu^2 \rangle + 2\alpha \langle V^\mu \zeta_\nu \rangle + \langle (V^\mu)^2 \rangle$$

(4.6)

is never negative. It is minimum for

$$\alpha = - \langle V^\mu \zeta_\nu \rangle / \langle \zeta_\nu^2 \rangle.$$

(4.7)

The statement that in particular the minimum is never negative yields the so-called *Schwarz inequality:*

$$\langle (V^\mu)^2 \rangle \langle \zeta_\nu^2 \rangle \geqslant \langle V^\mu \zeta_\nu \rangle^2,$$

(4.8)

which holds for any two real random quantities in any probability distribution.

The uncertainty relation. In our case, the right hand side is δ^μ_ν. According to a very general convention, we designate the square root of the variance of a fluctuation V by

$$\Delta V = \langle V^2 \rangle^{1/2}$$

(4.9)

and write the *uncertainty relation of statistics* in the form

$$\Delta V^\mu \Delta \zeta_\nu \geqslant \delta^\mu_\nu.$$

(4.10)

In particular we obtain for energy U and temperature T

$$\Delta U \Delta \frac{1}{T} \geqslant k,$$

(4.11)

for volume V and pressure p

$$\Delta V \Delta \frac{p}{T} \geqslant k,$$

(4.12)

and for particle number N and chemical potential μ

$$\Delta N \Delta \frac{\mu}{T} \geqslant k.$$

(4.13)

It is remarkable that these relations are independent of the size of the system, whereas the extensities U, V, N increase with the size.

Eq. (4.10) is an analogue to the uncertainty relation between complementary quantities in quantum mechanics. This analogy fails, however, in other respects. Different eigenstates of an observable in quantum mechanics are orthogonal to each other because the observables are *Hermitian* operators. By a measurement in an eigenstate we can find only the corresponding eigenvalue of the observable. In thermodynamics, however, ζ_ν is not represented by an *Hermitian* operator and therefore there is no "orthogonality" relation between different generalized canonical distributions with different values of ζ_ν. On the contrary, as was already pointed out, with a certain probability it is possible, for instance, to observe a certain temperature in an equilibrium with another temperature.

It should be stressed that the definition of the intensities ζ_ν by eq. (4.5) is not in agreement with the definition of temperature of a microcanonical distribution of eq. (4.27) in chapter 2.2.4. The temperature defined there is uniquely connected with energy. It is sharp as energy and is not a fluctuating quantity anyway; therefore it is not "complementary" to energy. The correct complementary intensity ζ_ν of eq. (4.5) is totally indetermined in the microcanonical distribution.

Minimum of uncertainty. We ask which distribution $W(V)$ corresponds to the equality sign in eq. (4.10). For simplicity we discuss the case of only one variable V. The inequality sign in eq. (4.10) means

$$\int dV(V + \alpha\zeta)^2 W = 0. \tag{4.14}$$

As the integrand is never negative, with eq. (4.5) this leads us to

$$(V + \alpha\zeta)W = \left(V + \alpha\frac{\partial}{\partial V}\right)W = 0. \tag{4.15}$$

This is a differential equation for $W(V)$ with the solution

$$W(V) = C\exp\left(-\frac{V^2}{2\alpha}\right). \tag{4.16}$$

It is a normalizable distribution only if α is positive. So we obtain the result that the normal distribution, which is distinguished in different respects, is also distinguished as the distribution with minimum uncertainty between the "complementary" quantities V and ζ.

5.3　Linear Thermodynamics

Nonequilibrium thermodynamics is the theory of changes with time of systems which can be described by thermal variables. It is a very large field in which limitation is not sharp. Only by certain restrictions it is possible to find laws and rules which are valid for certain classes of physical systems in this field. There is, however, a class of nonequilibrium processes for which laws are valid which in their generality are comparable with the laws of equilibrium thermodynamics. The field of these processes is called *linear thermodynamics* which will be defined in the following. This field is often called TIP (= thermodynamics of irreversible processes). Such a name, however, because of its literal meaning could be used for general nonequilibrium thermodynamics as well. Linear thermodynamics is concerned with processes which occur so in the neighborhood of a thermal equilibrium that it is allowed to linearize the macroscopic dynamical equations with respect to the quantities which describe the deviations from this equilibrium. The development of this part of thermodynamics is chiefly connected with the name of the Norwegian physicist *L. Onsager*. The last two chapters will demonstrate a method of *Einstein* in which an equilibrium will be interpreted as the equilibration of two competing processes. In one chapter it will lead us to an important connection between diffusion and friction, in the other chapter to the radiation formula of *Planck*.

5.3.1 Fluxes and Forces

Transport equations. We consider a deviation from a thermal equilibrium which is described with thermal variables which are zero in the equilibrium. These can be the set V of the deviations

$$V^\mu = M^\mu - M^{0\mu} \tag{1.1}$$

of extensive variables M^μ or alternatively the set ζ of the thermally conjugate intensities

$$\zeta_\nu = \lambda_\nu - \lambda_\nu^0 = \frac{1}{k} X_\nu. \tag{1.2}$$

The superscript zero designates the equilibrium values. As the thermal equilibrium is stable, these quantities will be a measure of the tendency to restore the equilibrium, as explained in connection with eq. (7.4) of chapter 2.3.7. In the framework of "linear thermodynamics" it is assumed that these deviations are always so small that a linear connection between these variables and their time derivatives can be posited. *L. Onsager* found out that the connecting equations assume universal properties if they connect the so-called thermodynamic *forces* X_ν with the so-called thermodynamic *fluxes* $\dot{V}^\mu$, which by convention are designated by I^μ, in the form

$$I^\mu = L^{\mu\nu} X_\nu. \tag{1.3}$$

The constant coefficients $L^{\mu\nu}$ are called *Onsager coefficients* ν is a summation dummy.

We shall analyse the assumptions contained in these equations in more detail. Generally dynamical equations which connect fluxes with the forces are called "transport equations". Linear thermodynamics is restricted to linear transport equations. A more general linear connection would be of the type

$$I^\mu(t) = \int\limits_0^\infty ds\, \Lambda^{\mu\nu}(s)\, X_\nu(t-s) \tag{1.4}$$

describing a "memory" of the system. This means that earlier forces X influence the present value I^μ. Indeed, such effects are observed in special materials. Then energy can be stored for a relatively longer time in internal degrees of freedom, for instance in molecular rotations. We exclude such cases and return to eq. (1.3) with the assumption that the time intervals s to which the "memory function" $\Lambda^{\mu\nu}$ contributes, lie on a short time scale distinctly separated from the macroscopic long time scale. Then we say that the system has *no memory*. Eq. (1.3) is applicable only on the long time scale. This means in particular that the time derivative in this equation has to be performed on the long time scale by a limiting process

$$I^\mu(t) = \lim_{\tau \to 0} \frac{1}{\tau} \{V^\mu(t + \tau) - V^\mu(t)\}, \tag{1.5}$$

where τ is large compared to the duration of microscopic processes but small on the macroscopic time scale, like in eq. (2.2) of chapter 5.2.2. This "macroscopic" time derivative is totally different from a derivative on the short time scale.

In statistical theory, the variables V^μ are mean values of phase functions in the Γ-space. Let us call them mean values on the "microscopic level". So far as occurring in eq. (1.3), they are conditional mean values performed at a time under the condition that the set X of the variables X_ν is given at an immediately preceding time. These conditional mean values have to be performed by use of a conditional probability which is progressive. This corresponds to the fact that thermodynamic probabilities generally are progressive, as explained in chapter 2.1.1. Accordingly eq. (1.3) is not invariant with respect to time reversal. They describe how the mean values V will develop, immediately after the X are given.

The *Onsager* hypothesis. If the system has "no memory", any small deviation from the equilibrium behaves independently of the way how it arose. It has to follow the same dynamics of eq. (1.3). This theorem is called the *Onsager hypothesis:*

> Small deviations from the equilibrium behave in the same way in time, independently of whether they are thermal fluctuations or whether they are prepared in any other way.

This hypothesis is a fundamental basis of irreversible thermodynamics. It allows us to transfer the dynamics of thermal fluctuations, which arose spontaneously, to deviations from the equilibrium which arose in a different way, say by a preparation with external devices.

The *Onsager* coefficients. We shall now consider thermal fluctuations in a thermal equilibrium. The macroscopically observable fluctuations V which are mean values on the microscopic level, and the corresponding X can be considered to be random quantities on a "mesoscopic level". Their mesoscopic mean value in the equilibrium is zero. The probability distribution on the "mesoscopic level" of the mesoscopic random quantities V is given by eq. (2.13) of chapter 5.1.2 yielding eq. (2.22) of the same chapter. We write this equation now in the form

$$\langle V^\sigma X_\nu \rangle = - k \delta_\nu^\sigma. \tag{1.6}$$

According to the *Onsager* hypothesis the mesoscopic random quantities V and X satisfy the transport equations eq. (1.3). Multiplying this equation by $V^\sigma(t)$ and performing the mean value in the thermal equilibrium with the distribution of eq. (2.13) of chapter 5.1.2 yields

$$L^{\mu\sigma} = -\frac{1}{k} \lim_{\tau \to 0} \frac{1}{\tau} \{\langle V^{\mu}(\tau) V^{\sigma}(0)\rangle - \langle V^{\mu}(0) V^{\sigma}(0)\rangle\}. \tag{1.7}$$

As the thermal equilibrium is invariant with respect to time shift, we were allowed to replace t with 0.

In accordance with the *Onsager* hypothesis, this result is valid not only for thermal fluctuations but for any small deviation from the equilibrium. It represents a remarkable connection between the *Onsager* coefficients in the phenomenological equation (1.3) and the time correlations of the thermal fluctuations V. It should be stressed that only if the transport equations (1.3) are expressed by the correct choice of the "fluxes" and "forces" as defined previously, the multipliers $L^{\mu\nu}$ are called *Onsager coefficients*. They are elements of the *Onsager matrix* L in the transport equations which in a short hand form can be written

$$I = LX. \tag{1.8}$$

5.3.2 *Onsager* Symmetry

We can divide the occurring V^{ν} into two classes depending on their behavior with respect to the time reflection which replaces the time t with $-t$. We symbolize the time reflection by an operator $\mathcal{T}$. According to the relation

$$\mathcal{T}^2 = 1, \tag{2.1}$$

which means that applying $\mathcal{T}$ twice yields the identity, the operator $\mathcal{T}$ has only the two eigenvalues $+1$ and -1. Thus

$$\mathcal{T} V^{\nu} = \epsilon(\nu) V^{\nu}, \qquad \epsilon(\nu) = \pm 1. \tag{2.2}$$

For instance, ϵ is $+1$ for the *Cartesian* coordinates $\vec{x}$ of ordinary space for velocities, momenta, or angular momenta it is -1. The thermal equilibrium is invariant with respect to a time reversal. This property is a consequence of the "microscopic reversibility" of the dynamics in the Γ-space. Therefore the equilibrium mean values in eq. (1.7) remain unchanged if all V are replaced with their time reversed

$$\langle V^{\mu}(t) V^{\nu}(0)\rangle = \epsilon(\mu) \epsilon(\nu) \langle V^{\nu}(0) V^{\mu}(-t)\rangle. \tag{2.3}$$

We are allowed to shift the time from $-t$ to 0 in the expression on the right hand side, and obtain

$$\langle V^{\nu}(0) V^{\mu}(-t)\rangle = \langle V^{\nu}(t) V^{\mu}(0)\rangle. \tag{2.4}$$

Thus the so-called *Onsager symmetry* is fulfilled:

$$L^{\mu\nu} = \epsilon(\mu) \epsilon(\nu) L^{\nu\mu}. \tag{2.5}$$

Onsager originally considered only values of ϵ equal to $+1$. *H. B. G. Casimir* paid attention to the occurrence of the value -1 and to the following correction which is necessary if in the system a constant magnetic field $\vec{B}$ is active, or if the system is rotating with a constant angular velocity $\vec{\omega}$. These two quantities have the eigenvalue -1 of ϵ. For $\vec{\omega}$ this is obvious. For $\vec{B}$ the time reversal means reversing all electrical currents which orignate the field. Therefore the microscopic reversibility is satisfied only if $\vec{\omega}$ and $\vec{B}$ are replaced with their negative.

$$L^{\mu\nu}(\vec{\omega}, \vec{B}) = \epsilon(\mu)\,\epsilon(\nu)\,L^{\nu\mu}(-\vec{\omega}, -\vec{B}). \tag{2.6}$$

These are the so-called *Onsager-Casimir relations*. They are fundamental in linear thermodynamics and are of the same rank as the Basic Laws of equilibrium thermodynamics. Their considerable importance will become more obvious with their application to special systems, which will be carried out in the chapters 5.3.4 and 5.3.5. The relations connect quantities which in the scope of the phenomenological description have very different physical origins. The symmetry was detected by *L. Onsager* already some years before World War II but it did not become known to a larger extent until the postwar years.

5.3.3 Entropy Production

Once again we consider a system in contact with a heat bath which is so large that it does not change the value of its intensity variables during the state changes of the system. We call this bath also the "environment". Let the state of the system be ρ^0 if this is in equilibrium with the environment. The momentary nonequilibrium state, however, shall be

$$\rho = \rho^0 \exp(\Xi - \zeta_\nu V^\nu). \tag{3.1}$$

As explained in chapter 2.3.6, the entropy production is never negative:

$$\Theta = -k\,\frac{\mathrm{d}}{\mathrm{d}t}\,K(\rho, \rho^0) = -k\dot{V}^\nu\,\frac{\partial K}{\partial V^\nu} \geqslant 0. \tag{3.2}$$

We remember that this quantity in general is not the increase of the whole entropy of the system, but the entropy produced in the interior of the system, and that it does not include the entropy which is exchanged with the environment. With eq. (1.2) of chapter 5.3.1 and eq. (3.15) of chapter 1.3.3 which we write in the form

$$\frac{\partial K}{\partial V^\nu} = -\zeta_\nu = -\frac{1}{k}\,X_\nu \tag{3.3}$$

and with

$$\dot{V}^\nu = I^\nu, \tag{3.4}$$

we obtain:

$$\Theta = I^\nu X_\nu \geqslant 0. \tag{3.5}$$

In the regime of linear thermodynamics where the transport equations

$$I^\mu = L^{\mu\nu} X_\nu \tag{3.6}$$

hold, entropy production is the quadratic form

$$\Theta = X_\mu L^{\mu\nu} X_\nu. \tag{3.7}$$

These two equations (3.6, 3.7) and the *Onsager* symmetry relations are the fundamentals of the *Onsager* theory of linear thermodynamics.

Vector flows. In the following we consider localizable quantities. $M^\nu(\vec{x})$ shall be local densities of extensive quantities and, corresponding to eq. (1.5), the variables $V^\nu(\vec{x})$ shall be the deviations of these densities from the equilibrium values. The flux $I^\nu(\vec{x})$ then is the time derivative of the density $M^\nu(\vec{x})$. In many cases, this is the density of a conserved quantity like for instance of energy, of mass, or of electrical charge. Then with a vector flow density $\vec{g}^\nu(\vec{x})$ of such a quantity, the continuity equation

$$I^\nu(\vec{x}) = - \vec{\nabla}\,\vec{g}^\nu(\vec{x}) \tag{3.8}$$

holds. But in general, for a not conserved quantity, we have to add a production rate $\omega^\nu(\vec{x})$:

$$I^\nu = - \vec{\nabla}\,\vec{g}^\nu + \omega^\nu. \tag{3.9}$$

($\vec{\nabla}$ is the Nabla operator $\partial/\partial\vec{x}$). To give an example, let M^ν be the density of the particle number of a certain chemical species in a mixture. $\vec{g}^\nu$ then is the particle flow density of the species. If chemical reactions do not take part, the particle number is a conserved quantity. If, however chemical reactions occur, ω^ν is the production rate density. This is the number of particles of the species which are produced by the reactions per unit of time and volume. The rate flux I^ν is the total increase of the particle number of the species per these units. The term with $\vec{g}^\nu$ corresponds to the exchange through the surface of the volume element. This may become more evident by the balance for a finite volume:

$$\frac{\mathrm{d}}{\mathrm{d}t}\int \mathrm{d}^3x\,M^\nu = -\int \mathrm{d}^2x\,g_n^\nu + \int \mathrm{d}^3x\,\omega^\nu. \tag{3.10}$$

The integral on the left hand side, as well as the second on the right one, are volume integrals. The first integral on the right hand side, however, is a surface integral. g_n^ν is the outward component of $\vec{g}^\nu$ normal to the surface.

Entropy production is

$$\Theta = \int \mathrm{d}^3x\,(- \vec{\nabla}\,\vec{g}^\nu + \omega^\nu) = \int \mathrm{d}^3x\,(\vec{g}^\nu\vec{\nabla}X_\nu + \omega^\nu X_\nu). \tag{3.11}$$

By comparison with eq. (3.5), this expression gives rise to the definition of the vector flows $\vec{g}^\nu$ and the production rates ω^ν as "fluxes", too. The corresponding "forces" then are $\vec{\nabla}X^\nu$ and X^ν respectively. Then we can comprise eqs. (3.5, 3.11) symbolically in the common form

$$\Theta = I \cdot X. \tag{3.12}$$

Now we restrict the discussion to a homogeneous substance or a homogeneous mixture and assume moreover that in the macroscopic description the fluxes I in a space point $\vec{x}$ are connected with forces X in the same point only. The connections then are called *local*. There the so-called *Curie principle* holds that fluxes I can be coupled in the transport equations only with forces X which are tensors of the same rank in ordinary space. Therefore the transport equations separate into

$$\omega^\mu = (L')^{\mu\nu}X_\nu, \tag{3.13}$$

$$\vec{g}^\mu = l^{\mu\nu}\vec{\nabla}X_\nu. \tag{3.14}$$

$l^{\mu\nu}$ is in ordinary $\vec{x}$-space a tensor of rank two. The following consideration will show that also the coefficients $l^{\mu\nu}$ fulfil the *Onsager-Casimir* relations with respect to the superscripts μ, ν.

On a short distance scale, eq. (3.14) should be written in the more detailed form

$$g_a^{\mu}(\vec{x}) = \int \mathrm{d}^3 x' \, \hat{l}_{ab}^{\mu\nu}(\vec{x} - \vec{x}') \, \frac{\partial}{\partial x_b'} \, X_{\nu}(\vec{x}'), \tag{3.15}$$

where the Roman subscripts designate the components in ordinary space. As the "non-locality" implicated in $\hat{l}(\vec{x} - \vec{x}')$ has only a short range, the differentiation can be shifted to $\hat{l}$ by partial integration. We thus obtain finally:

$$-\frac{\partial}{\partial x_a} \, g_a^{\mu}(\vec{x}) = \int \mathrm{d}^3 x' \left[\frac{\partial}{\partial x_a} \frac{\partial}{\partial x_b'} \, \hat{l}_{ab}^{\mu\nu}(\vec{x} - \vec{x}') \right] X_{\nu}(\vec{x}'). \tag{3.16}$$

On the scale of macroscopic distances. the kernel of the integral contrasts to a "local" kernel of singular type vanishing for $\vec{x} \neq \vec{x}'$ which yields a local relation

$$-\vec{\nabla}\vec{g}^{\mu} = (L'')^{\mu\nu}X_{\nu}, \tag{3.17}$$

in which all quantities depend only on the same space point $\vec{x}$. Comparison with eqs. (3.6, 3.8) shows that

$$L^{\mu\nu} = (L')^{\mu\nu} + (L'')^{\mu\nu}. \tag{3.18}$$

It shows moreover that L' and l in eqs. (3.13, 3.14) fulfil the same *Onsager-Casimir* relations as L.

Correct definition of fluxes and forces. Originally thermodynamic "fluxes" I were introduced as rate flows $\overset{\vee}{V}$ according to eq. (3.4). The corresponding forces X then can be defined as occurring in the bilinear form eq. (3.12) of the entropy production. Later on we extended the definition of "fluxes" with including the vector flows $\vec{g}$ and the production rate densities ω. The definition of the corresponding "forces" by the bilinear form of the entropy production can be maintained. They are the quantities X occurring in eq. (3.12) if this equation is used as the short hand form of eq. (3.11) in which the symbolic product of eq. (3.12) is specified.

5.3.4 Heat Conduction

As a special application of the *Onsager* theory, we consider the heat conduction in a homogeneous substance. This may, however, be anisotropic, that means that certain directions in ordinary space may be distinguished, like in a crystal. Heat conduction is a process in which no work occurs. All transported energy is heat. As internal energy is a conserved quantity, for its spatial density u the continuity equation

$$\dot{u} = -\vec{\nabla}\vec{g} \tag{4.1}$$

holds, where $\vec{g}$ is the flow density of heat. The thermal conjugate to u is $1/T$. According to eq. (1.2), the conjugate force is

$$X = \frac{1}{T} - \frac{1}{T^0} \, , \tag{4.2}$$

where T^0 is the equilibrium temprature. This yields

$$\Theta = \int d^3x \, \dot{u} X = \int d^3x \, \vec{g} \, \vec{\nabla} \, \frac{1}{T}. \tag{4.3}$$

It shows that each component of $\vec{\nabla}(1/T)$ is the "force" conjugate to the same component of the heat flow $\vec{g}$ as "flux". The corresponding transport equation (3.14) is

$$\vec{g} = l \, \vec{\nabla} \, \frac{1}{T}. \tag{4.4}$$

In an isotropic substance, l is a scalar, in an anisotropic crystal, however, it is a spatial tensor of rank two.

Positivity of entropy production shows that

$$\vec{g} \, \vec{\nabla} \, \frac{1}{T} = -\frac{1}{T^2} \, \vec{g} \, \vec{\nabla} \, \frac{1}{T} \geqslant 0. \tag{4.5}$$

This means that the heat flow can never go into a direction of increasing temperature. In particular, the heat conductive constant l of an isotropic medium always has to be positive.

The *Cartesian* components of $\vec{g}$ can be considered to be variables numbered by μ and ν. Then the *Onsager* symmetry yields that the tensor $l^{\mu\nu}$ has to be symmetric in space. This statement is trivial for crystals with a symmetry center, i.e. crystals symmetric with respect to the mirror reflection of all three coordinates x_ν into the opposite $-x_\nu$. It is, however, not trivial for crystals without a symmetry center which distinguish left from right hand orientation. Detailed measurements, in particular by *Ch. Soret* (1893) and *W. Voigt* (1903), showed that the heat conductive tensor $l^{\mu\nu}$ is indeed always symmetric. This was the historical starting point of the *Onsager* theory, by understanding that the symmetry is not based on the pure geometric symmetry of the crystal but has a fundamentally deeper origin in the properties of thermodynamic processes. This fundamental insight was the key to the generalization for all linear processes in thermodynamics.

5.3.5 Thermodiffusion

Another application of *Onsager* thermodynamics, the theory of *thermodiffusion*, will be discussed in this chapter. This is the coupling of heat transport with diffusion in a gas or a solution. We shall see that this process is a complex of several phenomena. We suppose that only one substance can diffuse. In the case of a gas we assume that this is a pure substance, in the case of a solution we assume that this is so dilute that the concentration of the solvent remains constant. We introduce again the deviations V of the extensities from the equilibrium values. V_1 shall be that of the energy density and V_2 that of the mole concentration. We change from particle to mole concentration to arrive at conventional quantities. The corresponding vector flows shall be $\vec{g}_1$ and $\vec{g}_2$.

The thermal conjugates are $1/T$ and $-\mu/T$ respectively, where μ is the molar chemical potential of the diffusing substance. Thus the transport equations (3.14) for this system are

$$\vec{g}_1 = l_{11}\,\vec{\nabla}\frac{1}{T} - l_{12}\,\vec{\nabla}\frac{\mu}{T}, \tag{5.1}$$

$$\vec{g}_2 = l_{21}\,\vec{\nabla}\frac{1}{T} - l_{22}\,\vec{\nabla}\frac{\mu}{T}. \tag{5.2}$$

The system is isotropic and therefore all $l_{\mu\nu}$ are scalars. The *Onsager* symmetry reads

$$l_{12} = l_{21}. \tag{5.3}$$

The *Onsager* coefficients $l_{\mu\nu}$ however, are not the directly observed quantities. Observed is not μ but p, the pressure of the gas, or the partial pressure of the dissolved substance in the solution. The general relation

$$d\mu = v\,dp - s\,dT \tag{5.4}$$

for a pure substance yields

$$\vec{\nabla}\mu = v\,\vec{\nabla}p - s\,\vec{\nabla}T, \tag{5.5}$$

where v is molar volume and s molar entropy of the diffusing substance. With molar enthalpy

$$j = \mu + Ts, \tag{5.6}$$

we can write

$$\vec{\nabla}\frac{\mu}{T} = j\,\vec{\nabla}\frac{1}{T} + \frac{v}{T}\,\vec{\nabla}p. \tag{5.7}$$

The transport equations take the form

$$\vec{g}_1 = (l_{11} - j\,l_{12})\vec{\nabla}\frac{1}{T} - l_{12}\,\frac{v}{T}\,\vec{\nabla}p, \tag{5.8}$$

$$\vec{g}_2 = (l_{21} - j\,l_{22})\vec{\nabla}\frac{1}{T} - l_{22}\,\frac{v}{T}\,\vec{\nabla}p. \tag{5.9}$$

In this form we shall discuss them for special processes.

The mechano-caloric effect. We consider the case that T is homogeneous and thus no temperature gradient is present:

$$\vec{g}_1 = -l_{12}\,\frac{v}{T}\,\vec{\nabla}p, \tag{5.10}$$

$$\vec{g}_2 = -l_{22}\,\frac{v}{T}\,\vec{\nabla}p. \tag{5.11}$$

The pressure gradient causes a matter flow $\vec{g}_2$ which carriers heat. This is the origin of a heat flow

$$\vec{g}_1 = \frac{l_{12}}{l_{22}}\,\vec{g}_2. \tag{5.12}$$

The thermo-molecular pressure difference. A state without diffusion, i.e. with matter flow $\vec{g}_2$ zero, can be maintained if the gradients of pressure and temperature assume adequate values satisfying

$$(l_{21} - j\, l_{22})\,\vec{\nabla}\,\frac{1}{T} = l_{22}\,\frac{v}{T}\,\vec{\nabla}p. \tag{5.14}$$

The corresponding ratio of the gradients is called the *thermo-molecular pressure difference:*

$$\frac{\mathrm{d}p}{\mathrm{d}T} = \frac{1}{vT}\left(j - \frac{l_{21}}{l_{22}}\right). \tag{5.15}$$

The *Onsager* symmetry now connects this ratio with the transport energy q:

$$\frac{\mathrm{d}p}{\mathrm{d}T} = \frac{j - q}{vT}. \tag{5.16}$$

In a solution, the pressure gradient has to be transformed into the concentration gradient. The possibility of maintaining the diffusion-free state by the adequate concentration gradient is called the *Ludwig-Soret effect*.

An opposite phenomenon, not discussed here, is that the diffusion of two substances into each other can produce a temperature gradient. This is called the *Dufour effect*.

Membranes. These effects are particularly important for semipermeable membranes which allow the solvent, but not the dissolved substance, to pass. If the membrane is rigid, no temperature gradient in the membrane but a pressure difference on both sides, the *osmotic pressure*, will appear.

If, however, the membrane is movable, a new situation has to be envisaged in the case that a second dissolved substance is present which can pass the membrane. The pressure difference on both sides will equilibrate to zero. This means that the partial pressure of the passing substance becomes different on both sides. This in return causes a temperature difference, the *osmotic temperature effect*.

5.3.6   The *Einstein* Relation of Diffusion

In this and the next chapter we shall discuss an ingeneous and important method of *A. Einstein* of connecting two quantities with each other which are associated with two different processes. This theory was formulated long before the development of the *Onsager* theory which is only indirectly related with that.

The relation between the diffusion flow $\vec{g}$ and the gradient of the mole density ρ in a solution is often written in the form of the *Fick law*

$$\vec{g} = -D\vec{\nabla}\rho. \tag{6.1}$$

The constant D is called the *diffusion coefficient*.

We point out that D is not an *Onsager* coefficient of type l_{22} of eq. (5.2) because the gradient of ρ is not a thermodynamic "force" in the correct definition of the *Onsager* theory. If no temperature gradient is present, eq. (5.9) reduces to

$$\vec{g} = -l_{22}\,\frac{v}{T}\,\vec{\nabla}p. \tag{6.2}$$

One mole of the diffusing substance carries the so-called *transport energy*

$$q = \frac{l_{12}}{l_{22}}. \tag{5.13}$$

As molar volume v is $1/\rho$, we may write

$$\vec{g} = - l_{22} \frac{1}{T\rho} \frac{\mathrm{d}p}{\mathrm{d}\rho} \vec{\nabla}\rho. \tag{6.3}$$

This is the form of eq. (6.1). It only yields a value D practically constant in space if the gradient of ρ is small enough.

The diffusion process can be described as the motion of particles on which a mechanical friction force originated by the solvent is acting. Let us consider the motion of such a particle on which moreover an external mechanical force with potential φ is acting. If $\vec{V}$ is the velocity and m the mass of the particle, the equation of motion

$$m\dot{\vec{V}} = - \vec{\nabla}\varphi - \frac{1}{B} \vec{V} \tag{6.4}$$

holds. B is called the *mobility* of the particle and gives a measure of the friction. It is assumed that after a negligibly short time the constant velocity

$$\vec{V} = - B\vec{\nabla}\varphi \tag{6.5}$$

of a stationary motion is reached.

Einstein established a relation between this velocity and the gradient of ρ in the following way. The thermal equilibrium in a potential φ described by the canonical distribution

$$\rho = C \exp(-\beta\varphi) \tag{6.7}$$

is interpreted to be the equilibrium of this motion with the opposite diffusion. With the "diffusion velocity" $\vec{V}_\mathrm{D}$ the relation

$$\vec{g} = \rho \vec{V}_\mathrm{D} \tag{8.8}$$

holds and eq. (6.1) can be written in the form

$$\vec{V}_\mathrm{D} = - D\vec{\nabla} \ln \rho = \beta D\vec{\nabla}\varphi. \tag{6.9}$$

This velocity has to be the opposite of $\vec{V}$ in eq. (6.5). In this way we obtain the so-called *Einstein relation*

$$D = kTB \tag{6.10}$$

between friction and diffusion. It is remarkable that temperature enters into this connection.

*5.3.7 *Einstein*'s Decution of *Planck*'s Radiation Formula

The method explained in the previous chapter of interpreting an equilibrium to be an equilibration of competing processes was also applied by *A. Einstein* to give a new derivation of *Planck*'s radiation formula. This derivation was remarkable in several respects. First, it already operated in an early state of quantum theory with the concept of photons. Moreover, it postulated for the first time the existence of the induced emission of light. This is a phenomenon which is of fundamental importance today for laser technology.

The two competing processes are the emission and the absorption of light by atoms. Due to the *Kirchhoff* law (chapter 3.3.2) the thermal equilibrium of the radiation is independent of individual properties of the atoms of the environment. For simplicity's sake we make the following assumptions: all these atoms shall be of the same kind occupying during the considered processes practically only two energy levels E_1 and E_2, which are connected with the frequency ν of light by

$$E_2 - E_1 = h\nu. \tag{7.1}$$

Einstein interpreted this already to be the energy of a "photon" and based his interpretation on the observations of the photo effect and its reversal, the X-ray production by electron impact. The existence of photons with a certain particle behavior was a new hypothesis. The number of atoms may be N_1 in the lower level E_1, and N_2 in the higher level E_2. The number of atoms which absorb a photon of energy $h\nu$ per unit time is assumed to be

$$\dot{N}_a = BN_1 f(T, \nu), \tag{7.2}$$

where $f(\nu, T)\, d\nu$ is the spatial energy density of the radiation at temperature T in the frequency band between ν and $\nu + d\nu$. The number of atoms which emit a photon of energy $h\nu$ per unit time is assumed to be

$$\dot{N}_e = [A + B'f(T, \nu)]\, N_2. \tag{7.3}$$

It consists of two parts. The first one is proportional to A which is the reciprocal of a finite life time of the excitation state of the atom due to *spontaneous emission*. The second term, however, proportional to f was a new assumption of *Einstein*. It corresponds to a *stimulated emission* of light which is induced by radiation of the same frequency and proportional to f. At that time the existence of the stimulated emission was another new hypothesis.

The equilibrium of the two processes means that

$$BN_1 f = (A + B'f)N_2 \tag{7.4}$$

and has to be the equilibrium

$$\frac{N_1}{N_2} = C \exp\left[\beta(E_2 - E_1)\right]. \tag{7.5}$$

This yields, with eq. (7.1),

$$f = \frac{a}{b \exp(\beta h\nu) - 1}. \tag{7.6}$$

The two constants a, b are determined by the following two requirements:

(1) For temperature increasing to infinity, which means for β going to zero, the energy density f has to become infinite as well. Therefore b is equal to one.

(2) For long waves, i.e. for low frequencies ν, classical physics has to be valid.

Classically, each standing wave has the mean energy kT of an oscillator. The number of standing waves in the frequency band between ν and $\nu + d\nu$ corresponding to eq. (6.5) of chapter 4.1.6 is given by

$$dz = \frac{8\pi\nu^2}{c^3}\, d\nu. \tag{7.7}$$

In the classical limit thus it holds that

$$f = \frac{8\pi\nu^2}{c^3}\, kT. \tag{7.8}$$

This is the *Rayleigh-Jeans formula* which was already known before *Planck*'s formula. It is correct only for low frequencies; it breaks down totally for higher frequencies where it predicts an infinitely increasing f. This unphysical divergence of classical theory is called the *ultraviolet catastrophy*. Eq. (7.8) yields

$$a = \frac{8\pi\nu^2}{c^3}\, h\nu. \tag{7.9}$$

Thus we regained *Planck*'s formula (eq. (6.10) of chapter 4.1.6):

$$f(T, \nu) = \frac{8\pi\nu^2}{c^3}\, \frac{h\nu}{\exp\left(h\nu/kT\right) - 1}\,. \tag{7.10}$$

This formula was already found experimentally by *W. Wien* in the regime of visual light before *Planck*'s quantum theory. It was in contradiction to any classical interpretation and was the motivation for *Planck*'s developing of his theory.

*5.4 A Model of Time Scale Separation

The time scale separation, the *Onsager separation*, is an important feature for a system to be describable by thermal variables in the usual way also in nonequilibrium. As already mentioned before, it is a very general assumption that most macroscopic systems have this feature. Nevertheless this property cannot be proved as a general consequence of the basic dynamics of a general class of systems. There is, however, a simple model for which this proof is possible and which will be discussed here. In the essentials but with certain modifications we shall follow a work of *R. J. Rubin*[25].

The model is a chain of equal mass points elastically coupled in which only one mass is far larger, say 10^9 times, than that of the other mass points. This heavy mass point is a simulation of a *Brownian* particle in the surroundings of molecules of a gas or a liquid. The dynamical variables of the heavy particle represent macroscopic quantities. A particular solution of the equations of motion and its behavior in the limit of large mass of the heavy particle will be discussed. We shall find that the system then shows the *Onsager* separation. This particular solution is remarkable because it

---

[25] *R. J. Rubin*, J. Math. Phys. 1, 309 (1960).

is identical with a thermal autocorrelation function of the heavy particle. This will be demonstrated in chapter 5.4.1. This fact, however, does not lie in line with the mentioned proof which is independent of statistics and which will be demonstrated in chapter 5.4.3. Beforehand, some mathematical tools will be made available in chapter 5.4.2.

*5.4.1   Autocorrelations in the Harmonic Chain

We consider a one-dimensional chain of mass points with masses coupled with their next neighbors by equal harmonic forces which could be realized by elastic springs. The deviation of the n-the mass point with mass m_n from its rest position may be described by the linear coordinate q_n. The momentum of this mass point is

$$p_n = m_n \dot{q}_n. \tag{1.1}$$

To avoid problems with exceptional situations at the two ends of the chain, we assume that the chain is closed, so that the mass point $n = N + 1$ is identical with $n = -N$. The *Hamiltonian* of the whole chain is

$$H(q,p) = \sum_n \left[\frac{1}{2m_n} p_n^2 + \frac{\kappa}{2} (q_n - q_{n-1})^2 \right]. \tag{1.2}$$

If we assume that the chain is in a thermal equilibrium with its environment of temperature T, the canonical distribution as a function of the variables (q, p) is a normal distribution. Therefore the following relations for the mean values hold at any time (see chapter 1.1.5):

$$\langle p_n p_l \rangle = kTm_n \delta_{nl}, \tag{1.3}$$

$$\langle p_n q_l \rangle = 0. \tag{1.4}$$

With

$$\dot{p}_n = -\frac{\partial H}{\partial q_n} = \kappa(q_{n+1} + q_{n-1} - 2q_n) \tag{1.5}$$

we thus obtain

$$\langle \dot{p}_n p_l \rangle = 0. \tag{1.6}$$

The expressions of eqs. (1.3, 1.4, 1.6) are thermal correlation functions of the variables at the same time.

Now we look at the autocorrelation function of p_n at different times:

$$\psi_n(t) = \langle p_n(t + t_0) p_n(t_0) \rangle = \langle p_n(t) p_n(0) \rangle. \tag{1.7}$$

In this equation we made use of the time shift invariance of any time correlation in the equilibrium. The function satisfies the same equation of motion as p_n. Due to eq. (1.3, 1.6), we find for $t = 0$ the initial values

$$\psi_n(0) = kTm_n, \tag{1.8}$$

$$\dot{\psi}_n(0) = 0. \tag{1.9}$$

ψ_n is symmetric in time by definition:

$$\psi_n(-t) = \psi_n(t). \tag{1.10}$$

A remarkable fact now is that for positive t the autocorrelation function ψ_n is identical with a particular solution p_n of the dynamical equations. We shall generally designate with $p_s(n, t)$ the particular solution $p_s(t)$ which is associated with the initial conditions

$$p_s(0) = \delta_{sn}, \tag{1.11}$$

$$\dot{p}_s(0) = 0. \tag{1.12}$$

This already involves

$$q_s(0) = 0. \tag{1.13}$$

For positive t thus the equation holds that

$$\psi_n(t) = kT m_n p_n(n, t). \tag{1.14}$$

The particular solution corresponds to the situation that the chain initially had been at rest till the n-th mass point was kicked at time zero. After this, the energy will gradually spread over the chain. Notwithstanding that for $t > 0$ the autocorrelation function ψ_n has the same time development, the solution $p_n(n, t)$ contains no statistical elements and is strictly deterministic. This point is important for the question of the origin of the time scale separation which already appears in the rigorous deterministic solution.

*5.4.2 Causal Functions

In this chapter some mathematical connections will be explained which will turn out to be useful for the study of the special dynamical solution defined in the previous chapter.

Any function $f(t)$ which is zero for all negative t is called a *causal function*. The *Fourier* transform

$$\hat{f}(\omega) = \int_0^\infty dt\, e^{i\omega t} f(t) \tag{2.1}$$

of a causal function has characteristic properties. If $\hat{f}(\omega)$ is regular, the analytical continuation

$$\hat{f}(\omega + i\eta) = \int_0^\infty dt\, e^{i\omega t - \eta t} f(t) \tag{2.2}$$

is obviously regular as well for all positive values η. The inversion of the *Fourier* transformation yields

$$f(t)e^{-\eta t} = \frac{1}{2\pi} \int_{-\infty}^\infty d\omega\, e^{-i\omega t} \hat{f}(\omega + i\eta). \tag{2.3}$$

With the complex variable

$$z = \omega + i\eta \tag{2.4}$$

this can be written in the form

$$f(t) = \frac{1}{2\pi} \int_{-\infty}^{\infty} d\omega\, e^{-izt} \hat{f}(z). \tag{2.5}$$

If the integration path is shifted only over regions of the complex z-plane where the integrand is regular, the integral remains unchanged. Therefore we can also write

$$f(t) = \frac{1}{2\pi} \int dz\, e^{-izt} \hat{f}(z), \tag{2.6}$$

where the path is an arbitrary parallel to the real axis of the z-plane and lying in the upper half of this plane, on which $\eta > 0$. All singularities of $\hat{f}(z)$ lie below such a path.

We consider only such causal functions which become zero for t going to plus infinity. Then partial integration yields

$$\int_0^{\infty} dt\, e^{izt} \dot{f}(t) = -iz \int_0^{\infty} dt\, e^{izt} f(t) - f(0). \tag{2.7}$$

Thus we obtain the rule for the *Fourier* transform of $\dot{f}(t)$:

$$\hat{\dot{f}}(z) = -iz\hat{f}(z) - f(0), \tag{2.8}$$

which further yields

$$\hat{\ddot{f}}(z) = -z^2\hat{f}(z) = izf(0) - \dot{f}(0). \tag{2.9}$$

These rules will be a convenient tool for the transformation of differential equations for a causal function into algebraic equations for the *Fourier* transform.

*5.4.3   The Macroscopic Motion

Now we specify the harmonic chain in order to obtain the simulation of a *Brownian* particle by putting all m_n equal to m which exception of the mass of this extraordinary particle

$$m_0 = \mu m, \tag{3.1}$$

where μ shall be considerably larger then 1. The equations of motion (1.15) lead us to

$$m\ddot{q}_n = -\kappa(2q_n - q_{n+1} - q_{n-1}) - (m_0 - m)\ddot{q}_0\delta_{n0}. \tag{3.2}$$

Collective modes. We introduce new variables describing a common motion of many mass points, so-called *collective modes*. First we write

$$\alpha = \frac{2\pi}{2N+1} k, \qquad (k = -N, \ldots +N). \tag{3.3}$$

Then we define the variables of the collective modes by

$$\gamma_\alpha(t) = \sum_{n=-N}^{N} e^{i\alpha n} q_n(t). \tag{3.5}$$

This is a kind of *Fourier* transformation with respect to the discrete variable n. There the "completeness relation" holds that

$$\frac{1}{2N+1} \sum_\alpha e^{i\alpha(n-l)} = \delta_{nl}, \tag{3.6}$$

with which we obtain the inverse transformation

$$q_n(t) = \frac{1}{2N+1} \sum_\alpha e^{-i\alpha n} \gamma(t). \tag{3.7}$$

Eq. (3.2) changes to a differential equation for the modes:

$$\ddot{\gamma}_\alpha + \left(w \sin\frac{\alpha}{2}\right)^2 \gamma_\alpha = -(\mu-1)\ddot{q}_0, \tag{3.8}$$

where we used the abbreviation

$$w = 2\left(\frac{\kappa}{m}\right)^{1/2}. \tag{3.9}$$

The initial conditions

$$q_n(0) = 0, \qquad p_n(0) = m_n \dot{q}_n(0) = \delta_{n0} \tag{3.10}$$

yield

$$\gamma_\alpha(0) = 0, \qquad \dot{\gamma}_\alpha(0) = \frac{1}{\mu m}. \tag{3.11}$$

Fourier **transforms.** All occurring functions of time are causal functions. With the rules of eqs. (2.8, 2.9) therefore, the *Fourier* transform of eq. (3.8) is

$$\left[-z^2 + \left(w \sin\frac{\alpha}{2}\right)^2\right] \hat{\gamma}_\alpha(z) - \frac{1}{\mu m} = (\mu-1)z^2 \hat{q}_0(z) + \frac{\mu-1}{m}. \tag{3.12}$$

This is an algebraic equation for $\hat{\gamma}_\alpha(z)$ with the solution

$$\hat{\gamma}_\alpha(z) = \frac{\dfrac{1}{m} + (\mu-1)z^2 \hat{q}_0(z)}{\left(w \sin\dfrac{\alpha}{2}\right)^2 - z^2}. \tag{3.13}$$

We are interested in the momentum $p_0(0, t)$ of the *Brownian* particle. For the sake of simplicity we designate it by

$$g(t) = \mu m \dot{q}_0(t), \tag{3.14}$$

which corresponds to

$$\hat{g}(z) = -i\,\mu m z \hat{q}_0(z).$$

(3.15)

On account of

$$q_0(t) = \frac{1}{2N+1} \sum_\alpha \gamma_\alpha(t),$$

(3.16)

this yields

$$\hat{g}(z) = -\mu m \frac{iz}{2N+1} \sum_\alpha \hat{\gamma}_\alpha(z).$$

(3.17)

With use of eq. (3.12) we obtain

$$\hat{g}(z) = [\mu + i(\mu - 1)z\hat{g}(z)]\xi(z),$$

(3.18)

where

$$\xi(z) = -\frac{iz}{2N+1} \sum_\alpha \frac{1}{\left(w \sin \frac{\alpha}{2}\right)^2 - z^2}.$$

(3.19)

Eq. (3.18) can be solved with respect to $\hat{g}$ yielding

$$\hat{g} = \frac{\mu \xi}{1 - i(\mu - 1)z\xi}.$$

(3.20)

$\xi(z)$ is the function $\hat{g}(z)$ for the special case of the homogeneous chain in which m_0 is equal to m. The singularities of $\xi(z)$ are the simple poles

$$\omega_k = w \sin \frac{\alpha}{2} = w \sin \frac{\pi k}{2N+1} \qquad (k = -N, \dots N)$$

(3.21)

lying on the real axis of the z-plane in the interval between $-w$ and w. They are the normal frequencies of the homogeneous chain.

The long chain. We consider now the case that the number $2N+1$ of mass points in the chain becomes very lage. We shall see that then the motion will become irreversible if we hold the observation time finite. This is not surprising if we keep in mind that large N are associated with long *Poincaré cycles* in which the system would return to its initial state. Cutting off the observation time at a fixed value means preventing the return.

For large N, the distance $\Delta\omega$ between neighboring frequencies ω_k becomes very small. We may, however, describe it as going to zero only if this difference is small compared to the reciprocal observation time. We know that, in the *Fourier* integral

$$g(t) = \frac{1}{2\pi} \int dz\, e^{-izt} \hat{g}(z)$$

(3.22)

of a causal function $g(t)$, the integration path has to be taken correctly, for instance parallel to the real axis of the z-plane in the upper half plane. We may, however, describe $\Delta\omega$ as going to zero if the distance η of the path from the real axis remains finite. The observation time t is, roughly speaking, on the scale of $1/\eta$. This means that we can describe $\Delta\omega$ as going to zero if the observation time remains finite. As *Poincaré* cycle frequencies are of the order of $\Delta\omega$, we see that this condition is the cutting off the time long before a cycle returns, which was mentioned above.

Under this condition of holding t finite, we may replace α with a continuous variable. With this we replace

$$\Delta\alpha = \frac{2\pi}{2N+1} \qquad \text{with } d\alpha \tag{3.23}$$

$$\frac{1}{2N+1} \sum_{\alpha} \dots \qquad \text{with } \frac{1}{2\pi} \int_{-\pi}^{\pi} d\alpha \dots . \tag{3.24}$$

The normal frequencies ω_k of the homogeneous chain thus are described as being continuously distributed over the interval of the real ω-axis between $-w$ and w. Then we obtain

$$\xi(z) = \frac{z}{2\pi i} \int_{-\pi}^{\pi} \frac{d\alpha}{\left(w \sin\frac{\alpha}{2}\right)^2 - z^2} . \tag{3.25}$$

It is convenient to transform the integration variable by putting

$$\operatorname{tg}\frac{\alpha}{2} = u, \tag{3.26}$$

which finally yields

$$\xi(z) = \frac{z}{w^2 - z^2} \frac{1}{2\pi i} \int_{-\infty}^{\infty} \frac{du}{(u-v)(u+v)} . \tag{3.27}$$

Here

$$v = \frac{z}{(w^2 - z^2)^{1/2}} \tag{3.28}$$

is uniquely defined by the requirement that the imaginary part of the root has to be positive. The integral in eq. (3.27) can be evaluated in the complex u-plane by shifting the integration path from the real axis over the upper half plane into plus infinity of the imaginary part. Then the integral remains the residuum of the integrand at the pole v. This yields

$$\xi(z) = \frac{z}{v(w^2 - z^2)} = \frac{1}{(w^2 - z^2)^{1/2}} . \tag{3.29}$$

If N increases unrestrictedly with finite observation time, the poles at the discrete normal frequencies ω_k will be changed into a cut along the interval from $-w$ to w of the real axis of the z-plane. In this cut two sheets of a twofold *Riemann* surface are connected. We obtain this cut also for arbitrary large times, if we first perform the limiting process to large N and subsequently that to large time t.

Up to now we discussed the singularities of $\xi(z)$. Now we shall discuss the singularities of $\hat{g}(z)$. This function possesses this cut as well, but has further singularities. They are roots of the equation

$$1 - i(\mu - 1)z\xi(z) = 0, \tag{3.30}$$

which is

$$i(\mu - 1)z = (w^2 - z^2)^{1/2}. \tag{3.31}$$

We always suppose $\mu > 2$. As $\hat{g}(z)$ is regular on the upper half of the z-plane, the pole

$$z_0 = -i\lambda \tag{3.32}$$

with positive

$$\lambda = \frac{w}{[\mu(\mu - 2)]^{1/2}} \tag{3.33}$$

is the only further singularity of $\hat{g}(z)$. Positivity of λ means that the pole z_0 lies on this *Riemann* sheet of the square root in eq. (3.31), on which the real part of this root is positive. We shall call it the "first" sheet.

Now we shall consider the evaluation of the integral eq. (3.22). We call to mind that the integration path can be taken parallel to the real axis if it lies in the upper half of the z-plane. For a path shifted into negative infinity of the imaginary part, the integral is zero. Therefore we can close the integral in the upper half plane by this path. In Figure 51 an already closed path is sketched which goes over into the required closed path by shifting the lower section of the path into minus infinity of the imaginary part of z. The shift does not change the integral so long as the shift does not pass singular points of $\hat{g}(z)$. Such singularities are only the winding points $\pm w$ and the pole z_0. We can shift the path of Figure 51 into the path of Figure 52 where the solid line lies on the first *Riemann* sheet. The broken line is the section of the path on the other sheet. A further part of the whole path is the circle running clockwise round the pole z_0 on the first sheet. We shall discuss the contributions of these different sections of the whole path.

The heavy particle. The circle round the pole gives us a term

$$g_h(t) = C\exp(-\lambda t). \tag{3.34}$$

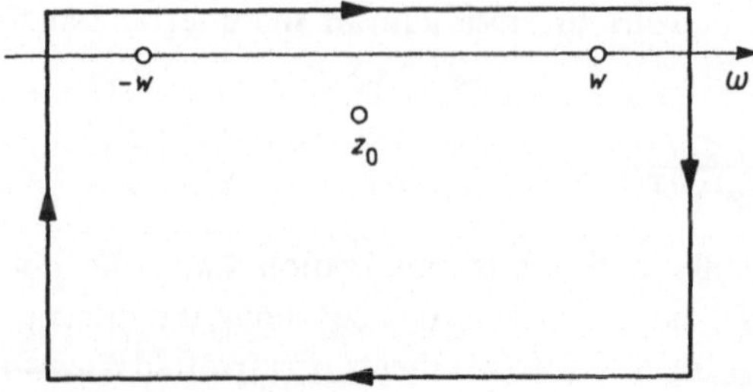

Figure 52

Integration path which arose with shifting the path of Figure 51. The solid line is the part of the path on one sheet and the broken line the part of the path on the other sheet of the *Riemann* plane.

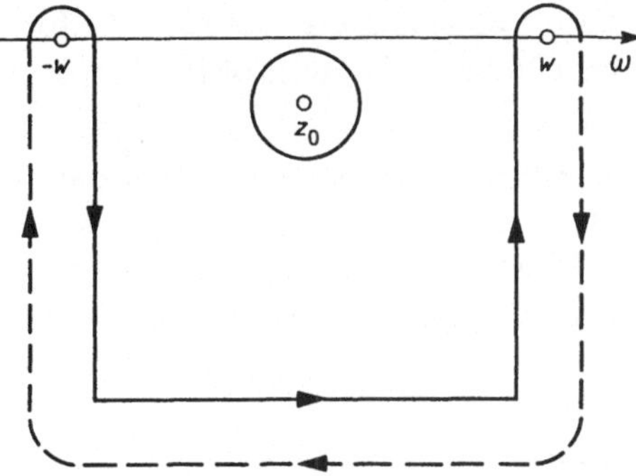

Figure 51

Integration path in the complex ω-plane. It circumscribes a pole at z_0 and a cut between $-w$ and w. This cut connects two sheets of a *Riemann* plane.

If μ is much larger than 1, we may write

$$\lambda = \frac{w}{\mu}. \tag{3.35}$$

The motion g_h then is very slow compared with oscillations of the frequency w. Therefore we call g_h the *macroscopic* or *hydrodynamic* part.

Now we come to the contribution $g_m(t)$ of the sections of the path which go parallel to the imaginary axis of z along the lines

$$z = \pm w - i\rho, \tag{3.36}$$

with ρ varying from zero to infinity. The values of $\hat{g}(z)$ on these sections of the path differ from each other only in the sign if they lie on both sheets at the same value z. As the path there, however has opposite directions, the contribution of both sections is two times that of the first sheet alone and we obtain

$$g_m(t) = e^{iwt}\frac{1}{\pi}\int_0^\infty d\rho\, e^{-\rho t}\hat{g}(-w-i\rho) - e^{-iwt}\frac{1}{\pi}\int_0^\infty d\rho\, e^{-\rho t}\hat{g}(w-i\rho). \tag{3.37}$$

This is a fast oscillation of frequency w modulated somehow by a decreasing amplitude. We call this term $g_m(t)$ the *microscopic* or the *short time* part of $g(t)$. Both motions g_h and g_m fade to zero with increasing time. The motion is thus irreversible. The time $1/\lambda$ is very large compared with the time $1/w$ of the fast oscillations if the mass m_0 of the *Brownian* particle is large.

We can note the following two essential results:

(a) The limiting process to large numbers of degrees of freedom with keeping the observation time fixed yields irreversibility.
(b) The limiting process to macroscopic variables yields the time scale separation.

We should like to stress that it is essential for obtaining the irreversibility for arbitrary large time that the limiting process to large numbers of degrees of freedom is performed before a limiting process to large time.

Long tail phenomena. A further remark may be made in passing. It is connected with a recent development in the study of large systems. Only values ρ contribute to the integral for which ρt is of order 1. Thus for t very large compared with $1/w$ only values ρ contribute which are very small compared with w. For such values ρ we may write

$$z^2 = w^2 \mp 2iw\rho. \tag{3.38}$$

We can rewrite eq. (3.20) for large μ in the form

$$\hat{g}(z) = \frac{(w^2 - z^2)^{1/2} + i\mu z}{w^2 - z^2 + \mu^2 z^2}, \tag{3.39}$$

which shows that the term proportional to z in the numerator does not contribute to g_m because it is cancelled on the two sections of the path on both sheets. Therefore we can drop the term in eq. (3.37). If moreover we take μ very large compared with 1, we may replace $\hat{g}_m$ by

$$\hat{g}_m(z) = \mu^{-1} w^{-2} (w^2 - z^2)^{1/2}, \tag{3.40}$$

$$\hat{g}_m(z) \approx \mu^{-1} w^{-2} (\mp 2iw\rho)^{1/2}. \tag{3.41}$$

Then eq. (3.37) yields

$$g_m(t) = \frac{\sqrt{2}}{\pi\mu} (\sqrt{i}\, e^{iwt} - \sqrt{-i}\, e^{-iwt}) (wt)^{-3/2} \int_0^\infty ds\, s^{1/2} e^{-s}. \tag{3.42}$$

This is again the oscillation with frequency w, however now with an amplitude which does not fade exponentially like $g_h(t)$, but with an algebraic exponent like $t^{-3/2}$. Notwithstanding that the microscopic motion g_m first remained in the background compared to the hydrodynamic motion g_h, it survives for a longer time and thus finally becomes the only motion.

Such "algebraic" regressions of autocorrelations were first observed by computer experiments and became in the last decade the object of detailed study. In this book, however, we will not be concerned with these phenomena in detail.

Generalizations. Up to now we have discussed the special model of the harmonic chain with the exceptional heavy particle. The results we are interested in with regard to macroscopic thermodynamics are the irreversibility and the time scale separation. We expect that these are features which are consequences of very general conditions for macroscopic systems. Yet in particular with respect to the time scale separation, the proof cannot be given in general. Nevertheless, after the study of the model, it is easier to depict this separation in the scope of a quantitative description in a more precise form than before, in the following way.

It is useful to consider autocorrelation functions of macroscopic variables. Each one is a function of time and can be represented by a *Fourier* transform which is a function of the complex *Fourier* variable z the complex "frequency". This function will have singularities in the z-plane. The singularities in the neighbourhood of the zero point of z correspond to slow motions, those far away from the zero point correspond to fast motions. The time scale separation is present if the singularities are distinctly separated into two groups. One group has sites near the zero point and represents the long time behavior, the second group are singularities far awy from the zero point and represent the short time behaviour.

Index

A

absolute equilibrium, *see* equilibrium
– temperature, *see* temperature
– zero point 114, 115
acceleration, gravitational, *see* gravitational acceleration
activation 116 f.
activity 93 ff., 113 f., 132
addition of knowledge 33 f.
adiabate (adiabatic) 81, 119
adjoint operator (*Hermitian* adjunction) 46
affinity 134 ff.
algebraic regression 240
alternative, simple 20
ammonia 132, 138
amorphous substance 111
angular momentum 179
antiferromagnets 185
a priori equal probability 27 f., 42
Arnold, V. A. 45
Arrhenius factor 116 f.
associative law of events 3
autocorrelations 232 f.
availability 19, 23, 90 ff., 134
available work 90
average, *see* mean value
Avez, A. 45

B

balance 134, 138
barometric pressure 141, 143, 149 ff., 203
Basic Laws 73 ff., 142
bath, *see* heat bath
Becker, R. 142
Berthelot 114
– bomb 94, 133
Bethe, H. 190
Bethe-Peierls approximation 190 ff.
bistochastic matrix 26
bit 20
bit-number 19 ff.
– in quantum theory 49
– cumulant 107 ff., 212
– variance 107 ff., 168
Bloch, F. 215
Bloch, see Wangsness-Bloch

Boer, see de Boer
Bohr, N. 217
Boltzmann, L. 145
Boltzmann equation 27
– constant 66, 203
–, *see Stefan-Boltzmann*
Bool algebra 4
Bosch, R. 138
Bosch, see Haber-Bosch
Bose attraction 165
– particle, *see* boson
– gas 147, 159 ff., 197 ff.
Bose-Einstein statistics 159 ff.
boson 54, 109, 147, 159 ff.
bound energy 94, 113
Boyle temperature 129
Bragg, W. L. 187
Bragg-Williams theory 187 ff., 192
Braun, see Le Chatelier-Braun
Brillouin, L. 205
Brillouin function 180 f.
Broglie, see de Broglie
Brownian particle 17, 231 ff.
Büchel, W. 205

C

caloric equation of state, *see* equation of state
– measurement 112, 134, 168
–, *see* mechano-caloric effect
calorimetric processes 62
canonical distribution, *see* distribution
carbon 138
Carnot, S. 71
Carnot cycle (engine) 70 ff., 76 ff.
Cartesian components 226
– coordinate 103, 148, 158, 222
Casimir, H. B. G. 220
Casimir, see Onsager-Casimir
causal function 233 f.
cell model 157 ff.
central limit theorem 17 f.
certainty 19, 21, 25
chain, harmonic 232
–, long 236
characteristic function 16
Chatelier, see Le Chatelier

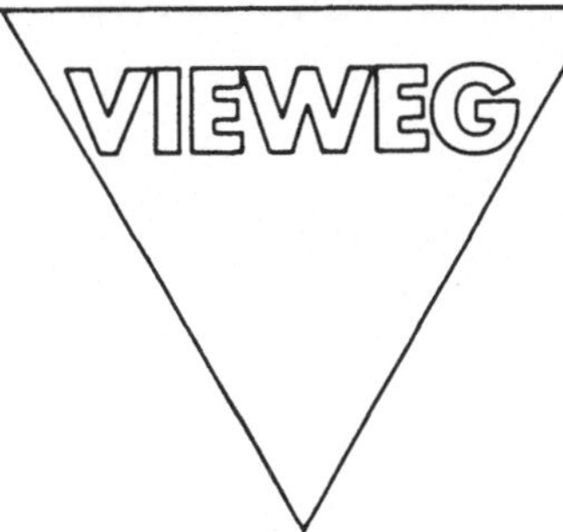

Festkörperprobleme
Advances in Solid State Physics

Editors: Fritz Sauter (Vols. I – V) – Otfried Madelung (Vols. VI – XII) – Hans J. Queisser (Vols. XIII – XV) – Joachim Treusch (Vols. XVI – XXI) – Peter Grosse (Vols. XXII – 27) – Ulrich Rössler (beginning Vol. 28)

According to the tradition of this series, every year a new volume is published. The forthcoming volumes provide up-to-date information by experts on solid state topics, which are in focus of interest. The volumes are addressed to senior scientists and engineers as well as to advanced students, who are interested in state of the art reviews and who look for an approach to the original literature.

Ulrich Rössler (Ed.)

Festkörperprobleme Advances in Solid State Physics 28

Plenary Lectures of the 52nd Annual Meeting of the German Physical Society (DPG) and of the Divisions "Semiconductor Physics" . . ., Karlsruhe, March 14 – 18, 1988.

1988. VI, 178 pages with 110 figures. 16,2 x 22,9 cm. Hardcover.

Volume 28 of Festkörperprobleme / Advances in Solid State Physics contains a selection of the plenary and invited talks of the spring meeting 1988 of the Condensed Matter Division (Arbeitskreis Festkörperphysik) of the German Physical Society.
The emphasis of the presented contributions is on current topics in semiconductor physics, which cover a broad spectrum from fundamental problems to applications and technology. This years winner of the Walter-Schottky Prize, M. Stutzmann, reports on electronic density of states and structural in hydrogenated amorphous silicon. Two contributions, devoted to the quantum-Hall effect, cover more recent investigations of this fundamental quantum phenomenon in solids. Theoretical aspects of semiconductors under high excitation conditions, the fabrication of the 4 Mega-bit random access memory, the physics of semiconductor microstructures under the aspects of reduced dimensionality, and the theory of high-T_c superconductors complete the list of topics.

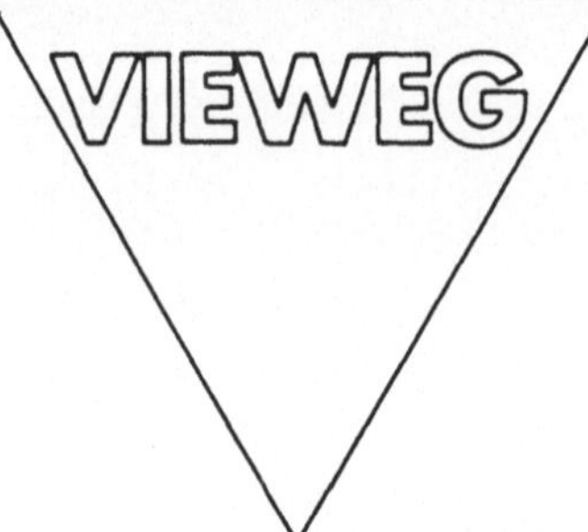

Harald Stumpf

Quantum Processes in Polar Semiconductors and Insulators

Volume 1
1983. XII, 388 pages. 16,2 x 22,9 cm. Hardcover.

Volume 2
1983. XII, 462 pages. 16,2 x 22,9 cm. Hardcover.

In these volumes the physics and the corresponding theory of processes and reactions of ideal and nonideal, i. e. impure polar semiconductors and insulators are discussed, resp. developed, in particular for I–VII and the technically interesting II–VI and III–V binary compounds. Based on the quantum theoretical microscopic description of crystals as many-particle systems of electrons and nuclei, a complete deduction is given from the microscopic level up to quantities which can be compared with experiment, i. e. average equilibrium and nonequilibrium values of quantum statistical processes and reactions with and without external fields.

More than 6.000 papers in this field were collected and taken into account, the greatest part of which is cited in these volumes. Due to the deductive character of the presentation it is possible to classify the various approaches in literature and to show their meaning within the framework based on recent developments by the author and scientist working with related problems. Thus a survey is given of the physics and theory in this field and, moreover, a systematic guide to the understanding of original papers.

Max Wutz, Hermann Adam and Wilhelm Walcher

Vacuum Engineering

Translated from the German by Walter Steckelmacher. 1989. Approx. 650 pages. 17 x 24,5 cm. Hardcover.

Contents: Introduction – Gaslaws, Foundation of the Kinetic Theory of Gases and Gasdynamics – Sorption and Desorption – Flow of Gases in Systems – Positive Desplacement Pumps – Fluid Entrainment Pumps – Molecular Pumps – Sorption Pumps – Condensers – Cryotechnology and Cryopumps – Vacuum Gauges and leak Detection Instruments – Leak Detection Technology – Materials – Vacuum Components and their Interconections to Vacuum Systems – Technology used dependent upon the different Pressure Regions – Appendix.